应用型本科计算机类专业"十三五"规划教材

# JAVA 程序设计实训教程

主 编 郑 豪 王 峥 王 洁
副主编 王小正 侯 青 朱 杰 李 青

南京大学出版社

## 图书在版编目(CIP)数据

JAVA 程序设计实训教程 / 郑豪，王峥，王洁主编
— 南京：南京大学出版社，2017.8 (2018.8重印)
应用型本科计算机类专业"十三五"规划教材
ISBN 978-7-305-19206-7

Ⅰ. ①J… Ⅱ. ①郑… ②王… ③王… Ⅲ. ①JAVA 语言－程序设计－高等学校－教材 Ⅳ. ①TP312.8

中国版本图书馆 CIP 数据核字(2017)第 193558 号

| | |
|---|---|
| 出版发行 | 南京大学出版社 |
| 社　　址 | 南京市汉口路 22 号　　邮编　210093 |
| 出 版 人 | 金鑫荣 |
| 丛 书 名 | 应用型本科计算机类专业"十三五"规划教材 |
| 书　　名 | JAVA 程序设计实训教程 |
| 主　　编 | 郑豪　王峥　王洁 |
| 责任编辑 | 王秉华　蔡文彬　　编辑热线 025-83597482 |
| 照　　排 | 南京理工大学资产经营有限公司 |
| 印　　刷 | 南京鸿图印务有限公司 |
| 开　　本 | 787×1092　1/16　印张 15.75　字数 404 千 |
| 版　　次 | 2017 年 8 月第 1 版　2018 年 8 月第 2 次印刷 |
| ISBN | 978-7-305-19206-7 |
| 定　　价 | 36.80 元 |

网　　址：http://www.njupco.com
官方微博：http://weibo.com/njupco
官方微信号：njupress
销售咨询热线：(025) 83594756

＊版权所有，侵权必究
＊凡购买南大版图书，如有印装质量问题，请与所购
　图书销售部门联系调换

# 前　言

　　Java 程序设计是计算机学科重要的编程语言之一，Java 程序设计实践教程也成为计算机学科最为重要的实践课程。为深化教育教学改革，进一步推动专业综合改革，依据国家十二五规划，瞄准 Java 发展前沿，面向经济社会发展需求，借鉴国内外教学改革成果，充分利用信息技术，由高校一线教师和知名企业工程师一起编写了这本 Java 程序设计实训教程。

　　本书以项目实战为主线，深入浅出，把 Java 的基础知识融入到各个实验的项目中，各个实验先以小项目训练读者基础知识，最后再以综合项目实战使用户深入理解。使读者真正的得到了项目实训，掌握了 Java 的程序开发，理解了 Java 的内在机制，也积累了项目开发经验。

　　全书分为 9 个实验，本书各章内容介绍如下：

　　实验 1 是初识 Java 程序设计，指导读者掌握 Java 的开发环境、基础语法等；

　　实验 2 是类的封装和继承，指导读者掌握类的继承性和封装性；

　　实验 3 是类的多态性，指导读者掌握类的多态性；

　　实验 4 是接口，指导读者掌握接口，并进一步深入了解多态性；

　　实验 5 是图形用户界面设计，指导读者能够用 swing 组件编写图形用户界面；

　　实验 6 是输入输出流，指导读者能够使用输入输出流对文件进行读写；

　　实验 7 是多线程设计，指导读者编写多线程的程序；

　　实验 8 是综合性实验 1，指导读者开发一个图像浏览器系统。

　　实验 9 是综合性实验 2，指导读者开发一个图书馆信息管理系统。

　　本书的所有实验代码均在 Eclipse 和 JDK 1.7 上通过编译和正常运行。实验 1 是由朱杰老师编写；实验 2 是由王洁老师编写；实验 3 是由王小正老师编写；实验 4，9 是郑豪老师编写；实验 5，6 是由侯青老师编写；实验 7 是由李青老师编写；实验 8 是由王峥老师编写。本书编写过程中参考了 Java 程序设计的相关文献，同时还查阅了大量的网络资料，在此对所有的作者表示感谢。

　　由于编者水平有限，书中不妥和错误之处还望读者批评指正。

编　者

2017 年 5 月

# 目　录

**实验 1　初识 Java 程序设计** ······· 1
1.1　知识点回顾 ······· 1
1.2　实验练习 ······· 4
1.3　项目实战 ······· 13
1.4　实验习题 ······· 16

**实验 2　类的封装和继承** ······· 18
2.1　知识点回顾 ······· 18
2.2　实验练习 ······· 24
2.3　项目实战 1 ······· 28
2.4　项目实战 2 ······· 30
2.5　实验习题 ······· 32

**实验 3　类的多态性** ······· 34
3.1　知识点回顾 ······· 34
3.2　实验练习 ······· 38
3.3　项目实战 1 ······· 43
3.4　项目实战 2 ······· 47
3.5　实验习题 ······· 51
3.6　小结 ······· 54

**实验 4　接口** ······· 56
4.1　知识点回顾 ······· 56
4.2　实验练习 ······· 58
4.3　项目实战 1 ······· 70
4.4　项目实战 2 ······· 73
4.5　实验习题 ······· 77

**实验 5　图形用户界面设计** ······· 81
5.1　知识点回顾 ······· 81
5.2　实验练习 ······· 93
5.3　项目实战 ······· 102
5.4　综合项目实战 ······· 107

  5.5 实验习题 …………………………………………………………………………… 144

## 实验 6 输入输出流(I/O) ……………………………………………………………… 146

  6.1 知识点回顾 ………………………………………………………………………… 146
  6.2 实验练习 …………………………………………………………………………… 150
  6.3 实验习题 …………………………………………………………………………… 156

## 实验 7 多线程 …………………………………………………………………………… 159

  7.1 知识点回顾 ………………………………………………………………………… 159
  7.2 实验练习 …………………………………………………………………………… 164
  7.3 项目实战 …………………………………………………………………………… 173
  7.4 实验习题 …………………………………………………………………………… 179

## 实验 8 综合项目 1——图像浏览器的实现 …………………………………………… 181

  8.1 项目概述 …………………………………………………………………………… 181
  8.2 项目需求分析 ……………………………………………………………………… 181
  8.3 总体设计 …………………………………………………………………………… 181
  8.4 项目文件结构说明 ………………………………………………………………… 183
  8.5 主要代码分析 ……………………………………………………………………… 184
  8.6 项目总结 …………………………………………………………………………… 200

## 实验 9 综合项目 2——图书馆信息管理系统 ……………………………………… 201

  9.1 项目概述 …………………………………………………………………………… 201
  9.2 项目需求分析 ……………………………………………………………………… 201
  9.3 项目数据库设计 …………………………………………………………………… 202
  9.4 项目总体设计 ……………………………………………………………………… 205
  9.5 主要代码分析 ……………………………………………………………………… 208

# 实验 1　初识 Java 程序设计

## 1.1　知识点回顾

1. Java 简介

Java 是一种可以撰写跨平台应用软件的面向对象的程序设计语言，是由美国 Sun Microsystems 公司于 1995 年 5 月开发的 Java 程序设计语言和 Java 平台的总称。Java 语言是从 C++程序语言发展而来的，但比 C++语言简单，它是当前比较流行的网络编程语言。Java 语言的出现是源于对独立于平台语言的需要，即这种语言编写的程序不会因为芯片的变化而发生无法运行或出现运行错误的情况。

Sun 公司对 Java 编程语言的解释是：Java 编程语言是个简单、面向对象、分布式、解释性、健壮、安全与系统无关、可移植、高效、多线程和动态的语言。主要特性如下：

(1) Java 语言简单易用。一方面，Java 语言的灵感主要来自于 C++语言，其语法与 C 语言和 C++语言接近，有 C++基础的开发者会感觉 Java 很熟悉，另一方面，Java 语言摒弃了 C++语言中很少使用的、难以理解且令人迷惑的那些特性，如操作符重载、多继承、自动的强制类型转换等。

(2) Java 语言是一种面向对象的程序设计语言。Java 语言只支持单继承，它的多继承是通过实现多接口来完成。

(3) Java 语言是分布式的。Java 拥有广泛的能轻易地处理 TCP/IP 协议的运行库，例如 HTTP 与 FTP 类库等等。这使得在 Java 中比在 C 或 C++中更容易建立网络连接。Java 应用程序可以借助 URL 通过网络开启和存取对象，就如同存取一个本地文件系统一样简单。

(4) Java 语言是健壮的。Java 语言的强类型机制、异常处理、内存垃圾的自动收集等是程序健壮性的重要保证；对指针的摒弃是 Java 语言的明智选择。

(5) Java 语言是安全的。

(6) Java 语言是平台无关的。所谓平台无关是指编译后的 Java 程序可直接在不同的平台上运行而不用重新编译，这一特性使得 Java 随着 Web 应用的普及而迅速普及起来。因此只要在操作系统中配有 JVM，就可以运行编译后的 Java 程序，也就是"一次编写，随处运行"，因而轻松实现跨平台。

(7) Java 语言是可移植的。Java 语言规定同一种数据类型在各种不同的实现中，必须占据相同的内存空间。

(8) Java 语言是解释型的。在运行时，Java 平台中的 Java 解释器对这些字节码进行解释执行，执行过程中需要的类在连接阶段被载入到运行环境中。

(9) Java 语言是高性能的。Java 语言是一种半编译半解释执行的语言。

(10) Java 语言是多线程的。在 Java 语言中，线程是一种特殊的对象，它由 Thread 类或子(孙)类来创建,或者实现 Runnable 接口来创建。Java 语言支持多个线程的同时执行,并提供多线程之间的同步机制来保证对共享数据的正确操作。

(11) Java 语言是动态的。Java 语言需要的类能够动态地被载入到运行环境,也可以通过网络来载入所需要的类。

2. Application 与 Applet

Application(应用程序)和 Applet(小应用程序)是 Java 提供的两种不同类型的程序。Java Application 是一种独立完整的程序,与其他应用程序类似,是可以在计算机操作系统中运行的程序;Java Applet 不是一种独立完整的程序,需要在浏览器这种特定环境下运行。

Java Application 程序的结构特点是:程序是由一个或多个文件组成的,每个文件又是由一个或多个类组成的,每个类是由若干个方法和变量组成的。

Java Applet 应用程序是内嵌于 HTML 文档中,并使用<APPLET>标记的可执行 Java 字节码,是具有动态、安全、跨平台特性的网络应用程序,通过主页发布到 Internet。Java Applet 可以在 Internet 中传输,通过因特网下载并且能在所有支持 Java 的浏览器中运行,它的最大特点是能对用户做出反应,并进行相应的变化。Applet 不能独立在 JVM 中运行,而是由浏览器或 Applet 阅读器(applet viewer)执行。

3. Java 开发运行环境介绍

Java 开发工具分为两大类:一类是基本开发工具,即 Sun 公司免费提供的 Java 2 SDK;另一类是专业开发工具,当今最流行的企业级开发工具有 Eclipse、MyEclipse、JBuilder2008、JDeveloper 等,功能强大,满足企业级开发的需要。

JDK(Java Development Toolkit)是 Sun 公司开发的 Java 开发工具包,它是一个简单的命令行工具,主要包括软件库、编译 Java 源代码的编译器、运行 Java 字节码的解释器,以及测试 Java Applet 的 Applet 阅读器,还有其他一些有用的工具。它主要是通过 DOS 命令行,在 DOS 环境下进行 Java 程序的编译和运行。可以到 Oracle 公司的 Java 语言官方网站上下载最新版的 JDK 软件。安装完 JDK 后,需要设置 3 个系统环境变量,JAVA_HOME 的路径、JDK 开发工具的路径和 CLASSPATH 的路径(具体配置见实验任务1)。

JCreator 是由 Xinox 软件公司所开发的一个可视化的 Java 程序集成开发环境,它给用户提供了包括工程管理、工程模板、代码实现、代码调试器、高亮语句编辑以及完全客户化的用户界面等广泛的功能。

JBuilder 是由 Borland 公司开发的一款功能强大的可视化 Java 集成开发环境,可以快速开发包括复杂企业级应用系统的各种 Java 程序,包括独立运行程序、Applet 程序、Servlet、JSP、EJB、Web Service 等。

Eclipse 是由 IBM 公司开发的一款开放源码的通用工具平台,它专注于为高度集成的工具开发提供一个全功能的、具有商业品质的工业平台,由 Eclipse 项目、Eclipse 工具项目、Eclipse 技术项目和 Eclipse Web 工具平台项目组成。最常用的 Java 开发功能实际上是 Eclipse 一个主要的插件 JDT(Java Development Tools,Java 开发工具)所提供的,它随 Eclipse SDK 一同发行。Eclipse 是开发源代码的项目,用户可以到 http://www.eclipse.org/downloads/去免费下载 Eclipse 的安装程序。图 1.1 所示的界面布局是在 Java 视图下的工作主界面。工作台是一个桌面开发环境,提供了使用 Eclipse 工具必需的用户界面,以 Java 视图为例,用户的工作界面包含窗口、菜单栏、工具栏等。在资源管理窗口中文件以"文件/目录"的形式进行管理,可以在该窗口中查找所需的文件资源并可以右键单击文件名进行相关操作。

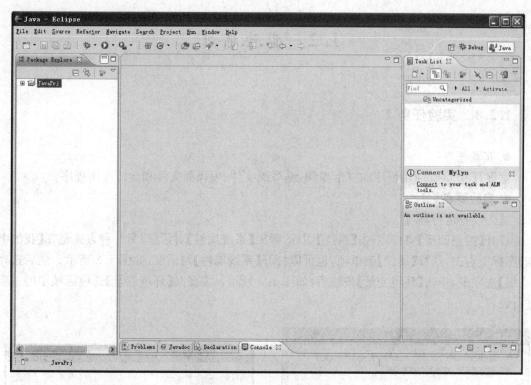

图 1.1 Eclipse 的工作界面

4. Java 核心包

(1) java.lang 包。封装所有编程应用的基本类，如 Object、Class、String、System、Integer、Thread 等。其中 Object 是所有类的根，它所包含的属性和方法被所有类继承。

(2) java.awt 包。封装抽象窗口工具包，提供构建和管理图形用户接口(GUI)设计工具，包含构件、容器和布局管理器。

(3) java.io 包。提供输入/输出文件操作的类。

(4) java.applet 包。为 Applet 提供执行所需的所有类。

(5) java.net 包。提供程序执行网络通信应用及 URL 处理的类。

(6) java.util 包。提供实用程序类和集合类，如系统特性定义和使用、日期方法类、集合 Collection、List、Arrays、Map 等常用工具类。

(7) java.sql 包。提供访问和处理标准数据源数据的类。

(8) java.rmi 包。提供程序远程方法调用所需的类。

(9) javax.naming 包。提供命名服务所需的类和接口。

(10) javax.swing 包。提供构建和管理程序的图形界面的轻量级的构件。

(11) javax.transaction 包。提供事物处理所需的基本类，除此之外，Javax 扩展包还提供 Rmi、Sound、Accessibility 等包。

## 1.2 实验练习

### 1.2.1 实验任务1

● 实验任务

掌握开发Java应用程序的三个步骤:编写源文件、编译源文件和运行应用程序。

● 实验要点

首先设置运行环境。

打开【控制面板】窗口,双击【系统】图标,弹出【系统属性】对话框;另一种方式是在【我的电脑】图标上右击,选择【属性】菜单项,也可以打开【系统属性】对话框,如图1.2所示。然后选择【高级】选项卡,单击【环境变量】按钮后,如图1.3所示。需要对【环境变量】选项区域中的选项进行设置。

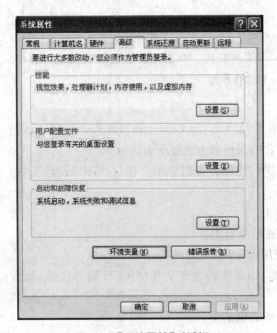

图1.2 【系统属性】对话框　　　　图1.3 【环境变量】对话框

① 设置 JAVA_HOME 环境变量

【系统变量】选项区域中通常没有JAVA_HOME变量,可以单击【新建】按钮,打开如图1.4所示的对话框。

在【变量名】文本框中输入JAVA_HOME,【变量值】文本框中输入安装路径,例如:C:\Program Files\Java\jdk1.6。

图 1.4 新建 JAVA_HOME 环境变量

② 设置 Path 环境变量

【系统变量】选项区域中通常已有 Path 变量（没有则新建），单击【编辑】按钮，打开如图 1.5 所示的对话框。

在【变量值】文本框中输入 Java 开发工具的所在路径，即 bin 文件夹的所在路径。可以填写绝对路径，例如 C:\Program Files\Java\jdk1.6\bin；或者输入相对路径，由于前面已经定义 JAVA_HOME 变量，所以相对路径是%JAVA_HOME%bin。

注意：【变量值】文本框中还有一些其他路径，每个路径之间用";"隔开。

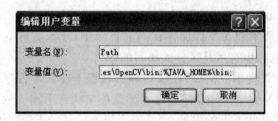

图 1.5 编辑 Path 环境变量

③ 设置 CLASSPATH 环境变量

【系统变量】选项区域中通常已有 CLASSPATH 变量（没有则新建），单击【编辑】按钮，打开如图 1.6 所示的对话框。

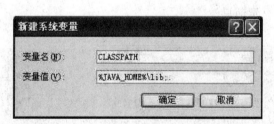

图 1.6 编辑 CLASSPATH 环境变量

Java 源程序一般用.java 作为扩展名，是一个文本文件，用 Java 语言写成，可以用任何文本编辑器来编辑。可以通过"程序"—>"附件"—>"记事本"来打开 Windows 自带的文本编辑器。程序编写完成后，保存文件。注意：文件名保存为 FirstJava.java。

使用 DOS 命令进入 FirstJava.java 文件所存放的目录（如 E:\java）下，使用 javac 命令编译该应用程序，通过编译该文件夹下会生成一个名为 FirstJava.class 的 class 文件，该文件包含程序的字节码，Java 字节码中包含的是 Java 解释程序将要执行的指令码，使用 java 命令执行该 class 文件，可在控制台下看到执行结果，如图 1.7 所示。

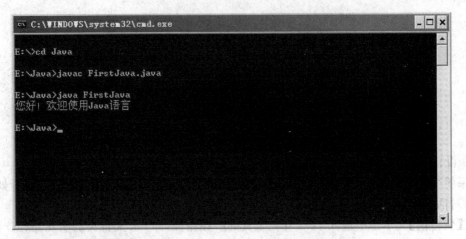

图 1.7　程序 FirstJava.java 在控制台下的输出结果

● **实验分析**

可能会遇到下列错误提示：

① Command not Found：没有设置好系统变量 Path。

② File not Found：没有将源文件保存在当前目录中，或源文件的名字不符合有关规定。要特别注意：Java 语言的标识符号是区分大小写的。

③ 出现一些语法错误提示。例如，在汉语输入状态下输入了程序中需要的语句符号等。Java 源程序中语句所涉及的小括号及标点符号都是英文状态下输入的，比如"你好，欢迎学习 Java"中的引号必须是英文状态下的引号，而字符串里面的符号不受汉语或英语的限制。

④ Exception in thread "main" java.lang.NoClassFoundError：没有设置好系统变量 ClassPath，或运行的不是主类的名字，又或程序没有主类。

● **实验主要代码**

```
/**
此类用于在屏幕上显示消息。
**/
public class FirstJava
{
    public static void main(String args[])
    {
        System.out.println("您好！欢迎使用Java语言");
    }
}
```

该程序的功能是在显示器屏幕的当前光标处输出显示字符串：你好！欢迎使用 Java 语言。

(1) 在 Java 中源程序的文件名并不是任意的，它必须和程序中定义的 public 类名相同，扩展名必须是 java。另外，由于 Java 是区分大小写的，所以也应确保文件名的大小写字母和类名一致。

(2) 所有的 Java 应用程序都通过调用 main() 方法开始执行，因此在 Application 的声明中对 main() 方法的定义是必不可少的。修饰符 static 说明该方法是静态的，它可以在创建对

(3) public class FirstJava 是一个类,是 Java 程序的基本组成部分。在 Java 中,所有功能都是以类的方式实现的。每一个类都是由关键字 class 和类名组成。其中类名是自己定义的,且含有 main()方法的类的类名和源程序文件的文件名必须一致。

(4) 可以添加注释。注释用来解释程序,使程序更好理解,它不会被执行。可以用//加单行的注释,用/*…*/或者/**…**/加多行的注释。

(5) 使用的变量名、对象名、方法名等标识符要有意义,尽量做到"见名知义"。

### 1.2.2 实验任务2

● 实验任务

使用 Eclipse 编写 Java 应用程序 FirstJava.java,根据命令行参数输出相应信息(如"您好!欢迎使用 Java 语言")。

● 实验要点

(1) 打开 Eclipse。双击 eclipse.exe,启动 Eclipse 平台。

(2) 配置 Eclipse 的 Workspace。初次打开 Eclipse,会要求用户配置 Eclipse 的 Workspace,点击 Browse,选择合适的工作区保存位置即可。

(3) 创建 Java 项目名 Java。执行 File—>New—>Java Project 菜单命令,打开 New Java Project 对话框。在 Project name 文本框中输入项目名称 Java,然后单击 Finish 按钮,如图 1.8 所示。

图 1.8　New Java Project 对话框

（4）创建Java类FirstJava。执行File->New->Class菜单命令，打开New Java Class对话框。在Name文本框中输入项目名称FirstJava，选中public static void main(String[] args)复选框，以自动创建main()方法框架，然后单击Finish按钮，如图1.9所示。

图1.9　New Java Class对话框

（5）在工作区编辑FirstJava代码。

（6）运行程序。执行Run->Run菜单命令，也可使用快捷键Ctrl+F11或单击调试工具栏中的启动运行按钮，运行结果如图1.10所示，提示用户输入参数。

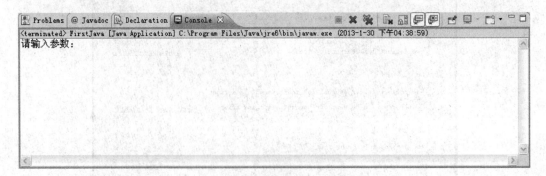

图1.10　提示用户输入参数

（7）配置运行参数并调试运行。执行Run->Run Configurations菜单命令，打开Run Configurations对话框。选择Arguments选项卡，在Program arguments文本框中输入命令

行参数"欢迎使用Java语言",如图 1.11 所示。

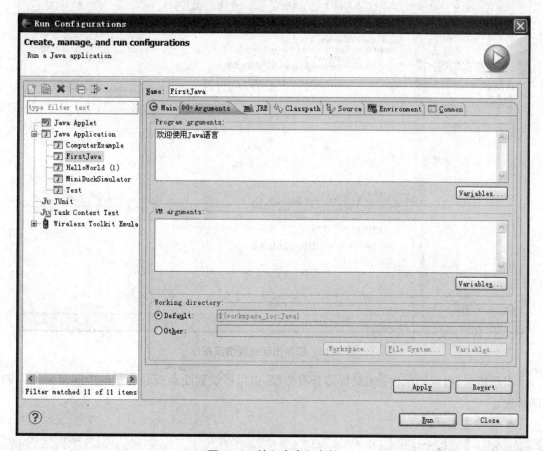

图 1.11　输入命令行参数

(8) 运行结果。单击 Run 按钮,运行结果如图 1.12 所示。

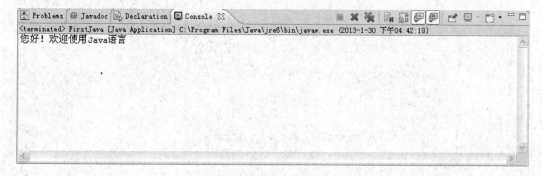

图 1.12　提供命令行参数的运行结果

● 实验分析

　　源代码编辑结束后,保存源代码,观察代码编辑窗口是否出现❌标志。如果在某行代码前出现该符号,说明这行代码存在语法错误,同时在控制台窗口会出现错误提示,需进行修改。比如在输入"System.out.println("请输入参数")语句时没有写";",就会出现如图 1.13 所示的错误提示。若没有出现该符号,说明源代码不存在语法错误,可以运行该程序。

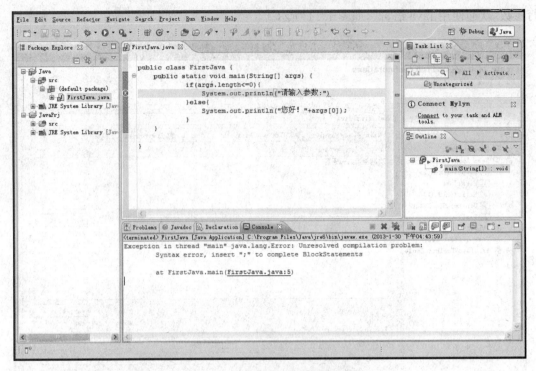

图 1.13　程序出现错误情况图示

出现错误，可以把鼠标停在❽标志查看错误，也可以设置断点查看变量的中间结果加以调试解决。

● 实验主要代码

```
public class FirstJava
{
    public static void main(String[ ] args)
    {
        if(args.length<=0)
        {
            System.out.println("请输入参数:");
        }else
        {
            System.out.println("您好!"+args[0]);
        }
    }
}
```

### 1.2.3　实验任务 3

● 实验任务

学习 Java 中 Java Applet 小程序的使用方法。

● **实验要点**

Applet 程序开发主要步骤如下：

1) 选用 Eclipse、记事本等工具作为编辑器建立 Java Applet 源程序。
2) 把 Applet 的源程序转换为字节码文件。
3) 编制使用 class 的 HTML 文件。在 HTML 文件内放入必要的<APPLET>语句。

下面举一个最简单的 Second 例子来说明 Applet 程序的开发过程：

（1）编辑 Applet 的 java 源文件

创建文件夹 E:\Java，在该文件夹下建立 Second.java

（2）编译 Applet

编译 Second.java 源文件可使用如下 JDK 命令：

E:\Java\javac Second.java<Enter>

注意：如果编写的源程序违反了 Java 编程语言的语法规则，Java 编译器将在屏幕上显示语法错误提示信息。源文件中必须不含任何语法错误，Java 编译器才能成功地把源程序转换为 appletviewer 和浏览器能够执行的字节码程序。

成功地编译 Second.java 之后生成响应的字节码文件 Second.class 的文件。用资源管理器或 DIR 命令列出目录列表，将会发现目录 E:\Java 中多了一个名为 Second.class 的文件。

（3）创建 HTML 文件

在运行创建的 Second.class 之前，还需创建一个 HTML 文件，appletviewer 或浏览器将通过该文件访问创建的 Applet。为运行 Second.class，需要创建名为 Second.html 的文件。

本例中，<APPLET>语句指明该 Applet 字节码类文件名和以像素为单位的窗口的尺寸。虽然这里 HTML 文件使用的文件名为 Second.html，它对应于 Second.java 的名字，但这种对应关系不是必须的，可以用其他的任何名字（比如说 temp.HTML）命名该 HTML 文件。但是使文件名保持一种对应关系可给文件的管理带来方便。

（4）执行 Second.html

如果用 appletviewer 运行 Second.html，需输入如下的命令行：

E:\java\appletviewer Second.html<ENTER>

可以看出，该命令启动了 appletviewer 并指明了 HTML 文件，该 HTML 文件中包含对应于 Second 的<APPLET>语句。

至此，一个 Applet 程序的开发运行整个过程结束了（包括 java 源文件、编译的 class 文件、html 文件以及用 appletviewer 或用浏览器运行）。

● **实验分析**

Applet 类是所有 Applet 应用的基类，所有的 Java 小应用程序都必须继承该类。

Applet 类的构造函数只有一种，即：public Applet()。

Applet 类中的四种基本方法用来控制其运行状态：init()、start()、stop()、destroy()。

（1）init()方法

这个方法主要是为 Applet 的正常运行做一些初始化工作。当一个 Applet 被系统调用时，系统首先调用的就是该方法。通常可以在该方法中完成从网页向 Applet 传递参数，添加用户界面的基本组件等操作。

（2）start()方法

系统在调用完 init()方法之后，将自动调用 start()方法。而且，每当用户离开包含该

Applet 的主页后又再返回时,系统又会再执行一遍 start()方法。这就意味着 start()方法可以被多次执行,而不像 init()方法。因此,可把只希望执行一遍的代码放在 init()方法中。可以在 start()方法中开始一个线程,如继续一个动画、声音等。

（3）stop()方法

这个方法在用户离开 Applet 所在页面时执行,因此,它也是可以被多次执行的。它使你可以在用户并不注意 Applet 的时候,停止一些耗用系统资源的工作以免影响系统的运行速度,且并不需要人为地去调用该方法。如果 Applet 中不包含动画、声音等程序,通常也不必实现该方法。

（4）destroy()方法

与对象的 finalize()方法不同,Java 在浏览器关闭的时候才调用该方法。Applet 是嵌在 HTML 文件中的,所以 destroy()方法不关心何时 Applet 被关闭,它在浏览器关闭的时候自动执行。在 destroy()方法中一般可以要求收回占用的非内存独立资源。(如果在 Applet 仍在运行时浏览器被关闭,系统将先执行 stop()方法,再执行 destroy()方法。

● 实验主要代码

```
import java.awt. * ;
import java.applet. * ;
public class Second extends Applet
{
    String str;
    public void init()
    {
        str = "Here is an Applet programme";
    }
    public void paint(Graphics g){
        g.drawString(str,100,100) ;
    }
}
```

将 Second.java 编译成字节码文件 Second.class。然后把文件 Second.class 嵌入到 Second.html 文件中,文件代码如下:

```
<HTML>
<BODY>
<APPLET CODE = "Second. class" width = 300 height = 300>
</APPLET>
</BODY>
</HTML>
```

在 Web 浏览器,运行 Second.html 文件,运行结果如图 1.14 所示。

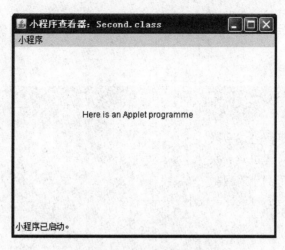

图 1.14　Applet 程序运行结果

## 1.3　项目实战

### 1.3.1　项目描述

公元前 5 世纪，我国古代数学家张丘建在他的《算经》中提出了著名的百鸡百钱问题。

有一人去买鸡，公鸡每只 5 元，母鸡每只 3 元，小鸡 3 只 1 元，用 100 元钱去买 100 只鸡，公鸡、母鸡、小鸡应该各买多少？

### 1.3.2　项目分析

根据问题中的约束条件将可能的情况一一列举出来，但如果情况很多，排除一些明显的不合理的情况，尽可能减少问题可能解的列举数目，然后找出满足问题条件的解。

完成百鸡百钱问题的有常规算法（懒惰枚举）和改进算法（非懒惰枚举）两种算法设计，不同算法对于同一问题可以有不同的枚举范围，不同的枚举对象解决问题的效益差别会很大。

(1) 算法设计一

懒惰枚举法：首先问题有三种不同的鸡，那么我们可以设公鸡为 x 只，母鸡为 y 只，小鸡为 z 只。由题意给出一共要用 100 钱买一百只鸡，如果我们全部买公鸡最多可以买 $100/5=20$ 只，显然 x 的取值范围是 $1\sim20$ 之间；如果全部买母鸡最多可以买 $100/3=33$ 只，显然 y 的取值范围在 $1\sim33$ 之间；如果全部买小鸡最多可以买 $100*3=300$ 只，可是题目规定是买 100 只，所以 z 的取值范围是 $1\sim100$。那么约束条件为：$x+y+z=100$ 且 $5*x+3*y+z/3=100$。

(2) 算法设计二

非懒惰枚举法：假如我设了公鸡和母鸡的个数为 x 和 y 了，那么公鸡和母鸡的数量就是确定的，那么小鸡的数量就是固定的为 $100-x-y$，那么此时就不再需要进行枚举了，约束条件就只有一个了：$5*x+3*y+z/3=100$。

### 1.3.3 项目编写

**算法设计一**

```java
public class ChickenBuy
{
    static int x;//可买公鸡只数
    static int y;//可买母鸡只数
    static int z;//可买小鸡只数
    public static void  lazyMethod()
    {
        x = 0;
        while(x<=19)
        {
            y = 0;
            while(y<=33)
            {
                z = 100 - x - y;
                if(x*5+y*3+z/3 == 100&&z%3 == 0&&x! = 0)
                {
                    System.out.println("可买公鸡只数:" + x);
                    System.out.println("可买母鸡只数:" + y);
                    System.out.println("可买小鸡只数:" + z);
                    System.out.println("————————————");
                }
                y++;
            }
            x++;
        }
    }
    public static void main(String[] args)
    {
        lazyMethod();
    }
}
```

**算法设计二**

```java
public class ChickenBuy
{
    static int x;//可买公鸡只数
```

```java
    static int y;//可买母鸡只数
    static int z;//可买小鸡只数
    public static void  notlazyMethod()
    {
        for(int x = 1;x<= 100 /5;x++ )
        {
            for(int y = 0;y<= 33;y++ )
            {
                z = 100 - x - y;
                if(x*5 + y*3 + z /3 = = 100 && z%3 = = 0)
                {
                    System.out.println("可买公鸡只数:" + x);
                    System.out.println("可买母鸡只数:" + y);
                    System.out.println("可买小鸡只数:" + z);
                    System.out.println("——————————");
                }
            }
        }
    }
    public static void main(String[ ] args)
    {
        notlazyMethod();
    }
}
```

运行结果如图1.15所示。

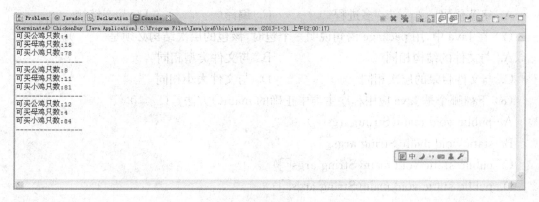

图1.15　百鸡问题求解结果

## 1.4　实验习题

1. 选择题

(1) 下列关于 Java 语言特性的描述中,错误的是(　　)。
A. 支持多线程操作
B. Java 程序与平台无关
C. Java 程序可以直接访问 Internet 上的对象
D. 支持单继承和多继承

(2) 下列关于 Java Application 程序结构特点的描述中,错误的是(　　)。
A. Java 程序是由一个或多个类组成的
B. 组成 Java 程序的若干个类可以放在一个文件中,也可以放在多个文件中
C. Java 程序的文件名要与某个类名相同
D. 组成 Java 程序的多个类中,有且仅有一个主类

(3) Java 程序经过编译后生成的文件的后缀是(　　)。
A. obj　　　　　B. exe　　　　　C. class　　　　　D. java

(4) 下列关于运行字节码文件的命令行参数的描述中,正确的是(　　)。
A. 第一个命令行参数(紧跟命令字的参数)被存放在 args[0]中
B. 第一个命令行参数被存放在 args[1]中
C. 命令行的命令字被放在 args[0]中
D. 数组 args[]的大小与命令行参数的个数无关

(5) Java 语言具有许多优点,下列选项中能反映 Java 程序的并行机制优点的是(　　)。
A. 安全性　　　　B. 跨平台　　　　C. 多线程　　　　D. 可移植

(6) 下列特点中,(　　)是 Java 虚拟机执行的特点之一。
A. 字节代码　　　B. 多进程　　　　C. 编译　　　　　D. 静态链接

(7) 在 Java 中,用 package 语句说明一个包时,该包的层次结构必须是(　　)。
A. 与文件的结构相同　　　　　　B. 与文件类型相同
C. 与文件目录的层次相同　　　　D. 与文件大小相同

(8) 下列哪个是 Java 应用程序主类中正确的 main()方法?(　　)
A. public void main(String args[])
B. static void main(String args[])
C. public static void main(String args[])
D. public static void main(String args[])

2. 编程题

(1) 编写一个阶乘应用程序。一个数 X 的阶乘(通常记作 X!)等于 X*(X−1)*(X−2)……*1。例如 4! 等于 4×3×2×1=24。创建一个应用程序,利用该应用程序可打印 2,4,6 和 10 的阶乘。

(2) 求解一个几何题程序。

已知一个直角三角形，其弦（最长边）的长度由下列公式给出：

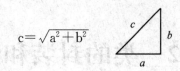

$$c=\sqrt{a^2+b^2}$$

编写一个 Java Applet 程序，从已知直角三角形的直角边计算最长边。

3. 问答题

(1) Java 语言有哪些特点？

(2) Java 语言程序分为哪两种？

(3) Java Application 编译后生成什么文件？该文件机器可以直接识别吗？

# 实验 2  类的封装和继承

## 2.1 知识点回顾

1. 类与对象

在面向对象程序设计(Object-Oriented Programming,OOP)中,使用"类"对现实世界的实体进行抽象和概括。类是组成 Java 程序的基本要素,通常封装同一类对象共同的属性和行为,例如人类具有姓名、国家、年龄等属性,也具有说话、吃饭、睡觉等行为。类是具有共同特性的一群实体的抽象,设计和定义类的过程就是将实体共有的属性和行为封装进类的过程,类中的属性被称为类的成员变量,行为被称为类的成员方法。

对象是现实世界中存在的一个实体,是由类创建的一个具体对象。如果说类是模板,对象则是由模板创建的一个实例,对象创建的过程也被称为实例化过程。由类可以创建出很多对象,例如由人类可以创建一个中国人、一个美国人、一个法国人,这些人类个体都是一个独立的对象,他们具有姓名、国家、年龄的属性,也具有说话、吃饭、睡觉的行为,这些对象均具有人类共同的属性和行为。所以类和对象的关系是总体和个体的关系,抽象和具体的关系。图 2.1 给出了类和对象的对比。

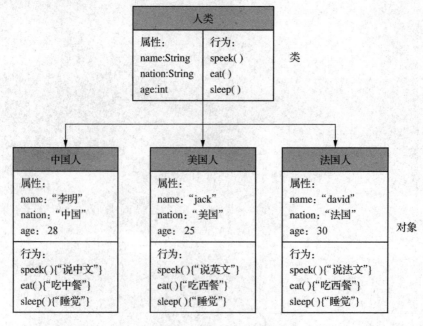

图 2.1  类和对象的对比

2. 类的定义和封装
(1) 类的定义
类定义的一般格式为：
［类修饰符］class 类名［extends 基类名］［implements 接口名］{
　　……　//成员变量声明 ⎫
　　　　　　　　　　　　　 ⎬类体
　　……　//成员方法声明 ⎭
}

class 关键字表示创建了一个类，extends 关键字表示该类继承了某一父类，implements 关键字表示实现了某接口。类的名字要符合 Java 标识符规定，即名字可以由字母、下划线、数字和 $ 符号组成，并且第一个字符不能是数字。给类命名时，通常要遵守下列编程习惯：① 类名通常以字母开头，首字母必须使用大写字母，如 Hello、Student、Book 等。② 类名要容易识别，一般可以由几个单词组合而成并且每个单词的首字母大写，如 HelloChina、BeijingTime、StudentInfo 等。

类体是类声明之后的一对大括号"{"、"}"以及它们之间的内容，主要描述该类共有的属性和行为，其中类的属性被成为成员变量，类的行为被成为成员方法。下面是一个类名为"Circle"的类，该类用来描述圆的属性（成员变量）和对圆的操作方法（成员方法），在类体的变量声明部分给出了 double 类型的成员变量 radius、area 分别表示圆的半径属性和面积属性，在方法定义部分给出了成员方法方法 calcArea()、setRadius()，分别用来计算圆面积和设置半径。

```
class Circle{
    double x, y, radius, area;           //成员变量声明部分
    double calcArea( ){
        area = 3.14159 * radius * radius;    //成员方法定义
        return area;
    }
    void setRadius(float r){             //成员方法定义
        radius = r;
    }
}
```

(2) 成员变量和成员方法
类体主要是对该类的成员变量进行声明以及成员方法进行定义，成员变量声明的一般格式为：
［修饰符］类型符 成员变量名［=初始值］;

成员变量名的类型可以是 Java 中任意的数据类型，既可以是 int、float、char 等简单类型，也可以是类、接口、数组等复杂数据类型。成员变量的初始值可以不赋值，如果不赋值成员变量将初始化为该类型的默认值。int、byte、short 类型初始值为 0，float、double 类型初始值为 0.0，char 类型初始值为空字符串，boolean 类型初始值为 false，string、数组及其他对象类型初始值为 null。

成员方法定义的一般格式为：
［修饰符］类型符 成员方法名（参数列表）{
　　……　//方法体
}

成员方法的类型符是该方法返回值的数据类型，如果该方法没有返回值则类型符为 void，如果有返回值则类型符为该返回值类型。参数列表是一组变量的声明，多个参数由逗号隔开，参数可以是任意的 Java 数据类型。

关于成员变量和成员方法的使用说明：

① 成员变量和成员方法的命名符合 Java 标识符规定，首字母使用小写字母，为了便于识别可以由几个单词组合而成，并且除第一个单词首字母小写外其后单词首字母大写，如变量 radius、方法 setRadius、方法 countArea 等。

② 成员变量和成员方法的修饰符有 public、protected、缺省和 private 四个级别，四个级别的使用权限不同。考虑到封装型的要求，通常成员变量声明为 private 则禁止外部直接访问，成员方法一般声明为 public 则提供给外部访问。四个修饰符的访问权限如表 2.1。

表 2.1　成员修饰符访问权限

| 访问级别 | 访问权限 |
| --- | --- |
| public | 公共的成员，访问不受限制，访问级别最高，可以被所有类访问 |
| protected | 受保护的成员，能被同包类和不同包子类访问 |
| 缺省 | 包可访问的成员，能被同包类访问 |
| private | 私有成员，访问级别最低，仅可以被本类访问 |

③ 类中除了有成员变量以外，还可以有局部变量。局部变量是指成员方法的参数或成员方法内部声明的变量。成员变量的作用范围是整个类，局部变量的作用范围只在声明它的方法内有效。当局部变量的名字和成员变量的名字相同时，成员变量会被隐藏，如果要使用被隐藏的成员变量，必须使用关键字 this，如：

```
class Circle{
    private double x , y , radius;              //声明成员变量 x, y, radius
    public void setRadius(double  radius){      //成员方法的参数 radius 为局部变量
        this.radius = radius;                   //局部变量 radius 传递给成员变量 radius
    }
}
```

(3) 对象的创建和使用

对象创建的过程包括了对象的声明、实例化和初始化 3 部分，对象创建的一般格式为：

类名　对象名＝new 类名(参数列表);

① 使用无参构造方法创建对象，如：Circle c1＝new Circle( );

其中 Circle c1 是对象的声明，c1 变量作为该对象的引用在后面的程序中使用，new Circle(.)是对象的实例化，使用 new 运算符为对象分配内存空间并返回一个引用，调用构造方法对对象进行初始化，在这里将调用无参构造方法进行初始化。

② 使用带参构造方法创建对象，如：Circle c2＝new Circle(2.0, 2.0, 1.0);

该条语句声明了圆类对象 c2，使用 new 运算符实例化圆类对象，并调用带参的构造方法进行该对象的初始化，并将对象的引用返回给 c2。

对象使用的一般格式为：

① 调用成员变量的格式：对象名.变量名

如：c1.radius=2.0;

② 调用成员方法的格式：对象名.方法名(参数列表);

如：c1.setRadius(2.0);

3. 构造方法及其重载

(1) 构造方法的定义和使用

构造方法是与类名同名，且不具有返回值的特殊函数，它的主要作用是进行对象的初始化工作，当对象被创建的时候会调用构造方法，构造方法定义的一般格式：

[修饰符] 构造方法名(参数列表){

……//构造方法体

}

关于构造方法定义和使用的说明：

① 构造方法必须和类同名。

② 构造方法没有返回值，它的工作是进行初始化。

③ 使用 new 运算符创建对象时构造方法即被调用，构造方法可以无参数也可以带参数，调用的时候会根据参数的情况来调用相应的构造方法。

④ 若类中没有定义构造方法，则 Java 虚拟机会提供一个默认的构造方法，该方法不带参数，方法体内无任何语句，什么工作也不做。

(2) 构造方法的重载

在一个类中，可以有多个构造方法，也可以没有构造方法。如果类中定义了多个方法，这些方法的名字相同，只是参数的类型和个数不同，这种情况被称为方法的重载。在定义构造方法时，通常会显式定义多个构造方法，这些构造方法的名字与类名相同，只是参数不同，这即是构造方法的重载。下面是关于构造方法重载的例子：

```java
class Circle{
    private double x , y , radius;
    Circle( ){                    //无参构造方法,对成员变量进行初始化
        x = 0.0;
        y = 0.0;
        radius = 1.0;
    }
    Circle(double x, double y, double radius){    //带参构造方法,对成员变量进行赋值
        this. x = x;
        this. y = y;
        this. radius = radius;
    }
}
```

在上面的例子中，Circle 类中定义了 2 个构造方法，分别是无参构造方法 Circle( )和带参构造方法 Circle(double x, double y, double radius)，这两个方法名字相同参数不同，构成了构造方法的重载。

## 4. 类的继承

计算机世界与自然界一样，类之间也有继承和派生关系。被继承的类叫父类或基类，继承的类叫子类或派生类。通过继承，子类可以获得父类的属性和行为（也被称为属性和方法）。继承与派生能达到代码重用、简化编程的目的。类继承的语法格式为：

　　［类修饰符］ class 子类 extends 父类{
　　　　……//类体
　　}

类的单一继承关系形成了清晰的层次结构，例如人类可以派生出学生类、教师类、工人类，学生类又可以派生出小学生类、中学生类、大学生类，教师类又可以派生出教授类、讲师类、助教类等。父类具有的属性和行为通过继承将会传递给子类，例如人类具有姓名、年龄等属性，具有吃饭睡觉等行为，学生类、教师类、工人类继承自人类，这些子类也将拥有姓名、年龄、吃饭、睡觉这些属性和行为。除了拥有人类共同的特征和行为，学生类还可以有学号、课程、学习、社团活动等子类独有的特征和行为。人类继承关系的树状结构如下图 2.2 所示。

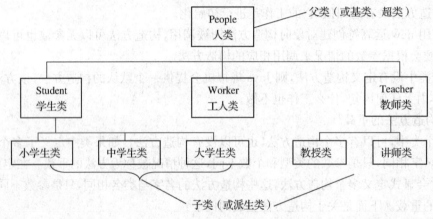

**图 2.2　人类继承关系的树状图**

在下面的例子中使用继承创建父类 People 和子类 Student，子类继承父类的属性和方法并定义自己的属性和方法。

```java
class People{
    String name;
    int age;
    public void eat( ){ …… }
    public void sleep( ){ …… }
}
class Student extends People{
    String no;
    Course course;
    public void learn( ){ …… }
}
```

在上面的例子中 Student 类继承自 People 类，Student 类被称为子类或派生类，People 类被称为父类或基类。子类除了继承了父类的属性和方法之外，还可以定义自己的属性和方法，所以 Student 除了具有属性 name、age，行为 eat( )、sleep( ) 以外，还具有 no 属性、course 属性、learn( ) 行为。

关于类的继承的说明：

① Java 中只支持单继承，也就是一个子类只能有一个父类，一个父类可以拥有多个子类。

② 继承具有传递性，子类沿继承路径向上继承所有父类的属性和方法。

③ Java 中的最顶层父类为 Object 类，如果一个类声明中没有指明这个类的直接父类，则默认该类继承自 Object 类。

④ 子类的构造方法与类名同名，在子类被创建的时候调用。在创建子类对象时，会首先调用父类的构造方法再调用子类的构造方法，如果没有显示调用父类的构造方法，则会调用父类的默认无参构造方法。

⑤ 可以在子类中使用 super 关键字显式调用父类的一个构造方法，但需要写在子类构造方法的第一行。

使用 super 关键字调用父类构造方法进行初始化的例子如下：

```java
class People{
    String name;
    int age;
    People( ){
        System.out.println("调用父类无参构造方法");
    }
    People(String name,int age){
        this.name = name;
        this.age = age;
        System.out.println("调用父类带参构造方法");
    }
}
class Student extends People{
    String no;
    Student(){
        System.out.println("调用子类无参构造方法");
    }
    Student(String name,int age,String no){
        super(name,age);              //显式调用父类带参构造方法
        this.no = no;
        System.out.println("调用子类带参构造方法");
    }
    public void printInfo( ){
        System.out.println("姓名" + name + ",年龄" + age + ",学号" + no);
    }
    public static void main(String args[ ]){
```

```
        //创建对象 s1 并调用无参构造初始化 s1
        Student s1 = new Student();
        s1.printInfo();
        //创建对象 s2 并调用带参构造初始化 s2
        Student s2 = new Student("李明",20,"2012001");
        s2.printInfo();
    }
}
```

上例中子类 Student 继承自父类 People,Student 继承了 People 类的 name、age 属性,并衍生了 no 属性。在创建子类对象 s1 时,会首先调用父类的无参构造方法再调用子类的无参构造方法,由于在无参构造里没有对 s1 进行任何初始化工作,所以输出 s1 的 name、age、no 属性时都将是 null 值。再创建子类对象 s2,s2 的初始化过程为先调用父类的构造方法对 name 和 age 进行初始化,再调用子类的构造方法对 no 进行初始化。在这个例子中父类的构造方法被显式调用了,即在 Student 子类构造方法通过 super(name,age)这条语句显式调用了父类 People 的构造方法,但需要注意的是这条语句必须放在 Student 子类构造方法的第一句以保证父类构造方法首先被调用。

## 2.2 实验练习

### 2.2.1 实验任务1

● 实验任务

编写程序定义圆类 Circle,把圆的性质和行为用代码块封装起来。圆的性质包括圆的半径和圆心坐标,行为包括计算圆面积,计算圆周长、设置半径、获取半径。实例化若干个圆对象,调用相应方法计算出圆面积、圆周长。

● 实验要点

(1) 掌握类的定义,抽象出类的属性和方法并封装进类中。

(2) 掌握类体的定义,了解构造方法的意义和作用,掌握构造方法的重载以及成员变量和成员方法的声明和定义。

(3) 掌握对象的实例化,有参构造方法和无参构造方法的定义和调用。

(4) 掌握使用 final 关键字声明符号常量。

● 实验分析

(1) 抽象出 Circle 类成员变量 radius、x 和 y,分别代表了圆的半径和坐标属性。成员方法 setRadius( )、getRadius( )、calcArea( )、calcGirth( ),分别实现设置半径、返回半径、计算面积、计算周长的行为。

(2) 使用 static final 关键字声明静态的符号常量 PI(符号常量通常定义为大写字母),并给其赋值为 3.14159,该常量可以被所有 Cirlce 类的圆对象使用,具体声明语句为:public static final double PI = 3.14159;

(3) 定义无参构造方法和带参构造方法,实现对圆对象的初始化。
(4) 编写程序的入口方法 main( ),在该方法中实例化若干个圆对象,并调用相应方法输出该圆的信息。

● **实验主要代码**

```java
public class Circle {
    private double radius;                                  //半径变量
    private double x, y;                                    //圆心坐标变量
    private static int num;                                 //圆对象个数变量
    public static final double PI = 3.14159;                //圆周率常量
    public Circle(){ num ++; }                              //无参构造方法
    public Circle(double r) throws Exception{               //带参数构造方法
        if (r < 0) { throw new Exception("圆半径不能为负数"); }
        else {radius = r;    num ++; }
    }
    public double getRadius(){  return radius;  }
    public void  setRadius(double r) throws Exception{
        if (r < 0) {
            throw new Exception("圆半径不能为负数");
        }else { radius = r; }
    }
    public static int getNum(){  return num; }
    public double calcArea(){  return PI * radius * radius; }
    public double calcGirth(){  return 2 * PI * radius; }

    public static void main(String args[]) throws Exception{
            Circle c1 = new Circle(3.5);
            System.out.println("圆1半径" + c1.getRadius() +",圆1面积" + c1.calcArea() +",圆1周长" + c1.calcGirth());
            System.out.println ("目前圆对象个数为" + Circle.getNum());
            Circle c2 = new Circle(10);
            System.out.println ("圆2半径" + c2.getRadius() +",圆2面积" + c2.calcArea() +",圆2周长" + c2.calcGirth());
            System.out.println ("目前圆对象个数为" + Circle.getNum());
            Circle c3 = new Circle();
            System.out.println ("圆3半径" + c3.getRadius() +",圆3面积" + c3.calcArea() +",圆3周长" + c3.calcGirth());
            System.out.println ("目前圆对象个数为" + Circle.getNum());
    }
}
```

程序运行结果如图 2.3 所示。

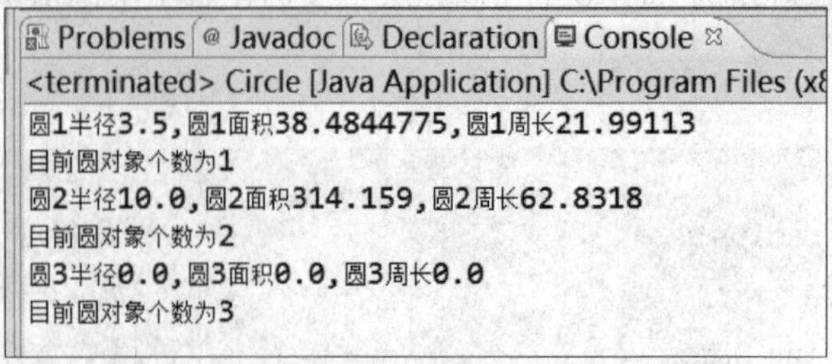

图 2.3　程序运行结果图

### 2.2.2　实验任务 2

● 实验任务

编程实现类的继承。编写父类 People，子类 Student 继承自 People。人类具有姓名，性别，年龄等属性，还具有吃和说的行为。学生类继承父类，除继承父类特征外还拥有学号属性和学习行为。构造人类和学生类的对象，调用方法输出有关信息。

● 实验要点

(1) 掌握类的定义，掌握构造方法以及成员变量和成员方法的定义和调用。
(2) 了解成员访问控制符的访问权限，掌握成员访问控制符的设置和应用。
(3) 理解类的继承的概念和含义，掌握类的继承的实现。
(4) 理解构造方法继承的概念和含义，掌握 super 关键字的使用。

● 实验分析

(1) 定义人类 People 为父类，学生类 Student 为子类。人类具有姓名 name、年龄 age 的属性及吃 eat( )行为和说 speak( )行为，学生也继承这些属性和行为并且另外添加学号属性 stuNo 和学习行为 learn( )。

(2) 成员属性封装为 private，表示成员属性私有仅可以被本类访问；成员方法定义为 public，表示该方法公有可以被所有类访问。

(3) 编写程序的入口方法 main( )，在该方法中实例化人类对象和学生对象，并调用对象的相应方法输出信息。

● 实验主要代码

```java
package ext;
//People 类
public class People {
    private String name;        //name、age 属性定义为 private 私有
    private int age;
    public People(){}                    //无参构造
    public People(String name, int age){    //带参构造
        this.name = name;
        this.age = age;
```

```java
    }
    public String getName() {         //获取学号的方法
        return name;
    }
    public void setName(String name) {      //设置姓名的方法
        this.name = name;
    }
    public int getAge() {     //获取年龄的方法
        return age;
    }
    public void setAge(int age) {     //设置年龄的方法
        this.age = age;
    }
    public void eat(){                            //吃方法
      System.out.println(name + "在吃饭");
    }
    public void speak(){                          //说方法
        System.out.println(name + "在说话");
    }
}

package ext;
//Student 类继承自 People 类
class Student extends People {
    private String stuNo;               //学号属性
    public Student(){}
    public Student(String stuNo, String name, int age){
            super(name, age);
            this.stuNo = stuNo;
    }
    public String getStuNo() {
        return stuNo;
    }
    public void setStuNo(String stuNo) {
        this.stuNo = stuNo;
    }
    public void learn(){
        System.out.println(super.getName() + "在学习");
                                    //使用 super 调用从父类获取姓名
```

```
    }
    public static void main(String[] args) {
        People p = new People("王浩宇", 30);
        p.eat();
        p.speek();
        System.out.println();                    //换行
        Student stu = new Student("001", "秦岭", 20);
        stu.eat();
        stu.speak();
        stu.learn();
    }
}
```

程序运行结果如图 2.4 所示。

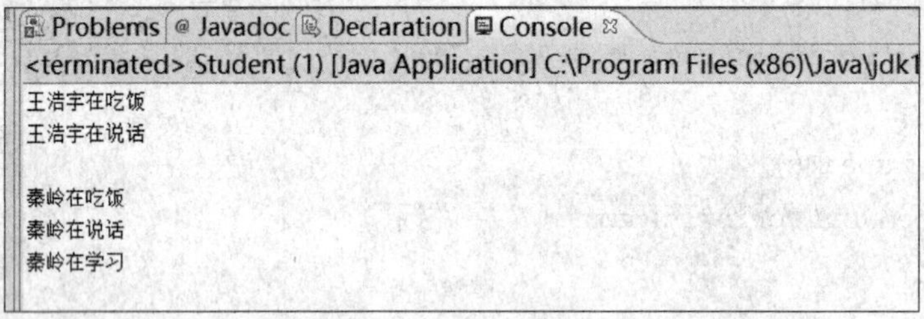

图 2.4　程序运行结果图

## 2.3　项目实战 1

使用 Java 的继承关系来描述动物世界的特征和关系。

### 2.3.1　项目描述

（1）抽象出项目问题中的类：动物、老鼠、熊猫。
（2）抽象出类属性：名字、食物；抽象出类行为：吃饭、睡觉、打洞（老鼠都有）。
（3）抽象出继承关系，老鼠类和熊猫类继承自动物类，子类继承父类的属性和方法。

### 2.3.2　项目分析

（1）动物世界的名字和食物是共有的属性，吃和睡觉是共有的行为。在 ext 包下创建 Animal 类，在该类中定义成员变量 name 和 food，成员方法 eat( )和 sleep( )。根据封装性的要求将 name 和 food 属性定义为 private 私有，其他类要访问该属性需通过 public 公有的 setter 和 getter 方法进行。
（2）创建老鼠类 Mouse 和熊猫类 Panda，这两类均继承自 Animal 类，另外老鼠类还具有打洞方法 dig( )。

（3）创建测试类 AnimalTest，编写程序入口 main( )方法，在该方法中创建老鼠和熊猫对象，调用相应方法输出结果。

### 2.3.3 项目编写

创建父类 Animal，代码如下：

```java
package ext;
public class Animal {
    private String name;
    private String food;
    public void eat() throws Exception{    //吃的行为
        System.out.println(this.name + "吃" + this.food + "!");    }
    public void sleep(){      //睡觉行为
        System.out.println(this.name + "在睡觉!");    }
    public String getName()
    {return name;    }
    public void setName(String name)
    {  this.name = name;    }
    public String getFood()
    {  return food;    }
    public void setFood(String food)
    {this.food = food;
    }
}
```

创建子类 Mouse，代码如下：

```java
package ext;
public class Mouse extends Animal {
    public Mouse() throws Exception{
        this.setName("老鼠");
        this.setFood("残羹冷炙!");
    }
    public void dig(){    // 打洞行为
        System.out.println(this.getName() + "会打洞");    }
}
```

创建子类 Panda，代码如下：

```java
package ext;
public class Panda  extends Animal {
```

```
    public Panda() throws Exception{
        this.setName("熊猫");
        this.setFood("竹子");
    }
}
```

创建测试类 AnimalTest,代码如下：

```
package ext;
public class AnimalTest {
    public static void main(String[] args) throws Exception   {
        Mouse m = new Mouse();      //创造老鼠
        m.eat();                    //老鼠吃
        m.sleep();                  //老鼠睡觉
        m.dig();                    //老鼠打洞
        System.out.println("--------分割线--------");
        Panda p = new Panda();      //创造熊猫
        p.eat();                    //熊猫吃
        p.sleep();                  //熊猫睡觉
    }
}
```

最后测试类 AnimalTest 的运行结果如图 2.5 所示。

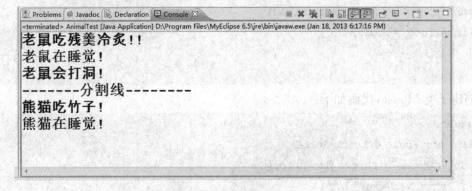

图 2.5　程序运行结果图

## 2.4　项目实战 2

使用 Java 的继承关系来描述计算机和笔记本计算机之间的特征和联系。

## 2.4.1 项目描述

(1) 抽象出项目问题中的类：计算机 PC、笔记本计算机 Notebook PC。
(2) 抽象出类的属性：CPU 主频、内存容量。抽象出类的行为：显示对象属性信息。
(3) 抽象出继承关系，Notebook PC 类继承自 PC 类。

## 2.4.2 项目分析

(1) 创建 PC 类，该类具有 CPU 主频和 RAM 内存容量的属性，以及输出属性信息的方法 showInfo()。
(2) 创建 Notebook PC 类，该类继承自 PC 类，使用无参构造方法和带参构造方法进行对象信息初始化，注意 super 关键字的显式调用使用。
(3) 创建测试类 PCTest，编写程序入口 main( ) 方法，在该方法中创建笔记本对象，调用相应方法输出结果。

## 2.4.3 项目编写

创建父类 PC，代码如下：

```java
package ext;
public class PC {
    private String cpu;    //私有成员属性cpu
    private String ram;    //私有成员属性 ram
    public PC(){}          //无参构造方法
    public PC(String cpu,String ram){  //带参构造方法
        this.cpu = cpu;
        this.ram = ram;
    }
    public void showInfo(){         //公有成员方法 showInfo
        System.out.println("这台计算机主频是:" + cpu + ",内存容量是:" + ram);
    }
}
```

创建子类 Notebook PC，代码如下：

```java
package ext;
public class NotebookPC extends PC {
    public NotebookPC(){ }   //调用父类无参构造方法
    public NotebookPC(String cpu,String ram){
        super(cpu,ram);      //调用父类带参构造方法
    }
}
```

创建测试类 PCTest，代码如下：

```java
package ext;
public class PCTest {
    public static void main(String[] args) {
        // TODO Auto-generated method stub
        NotebookPC npc = new NotebookPC("4GHZ","64G");
        npc.showInfo();
    }
}
```

最后，测试类 PCTest 的运行结果如图 2.6 所示。

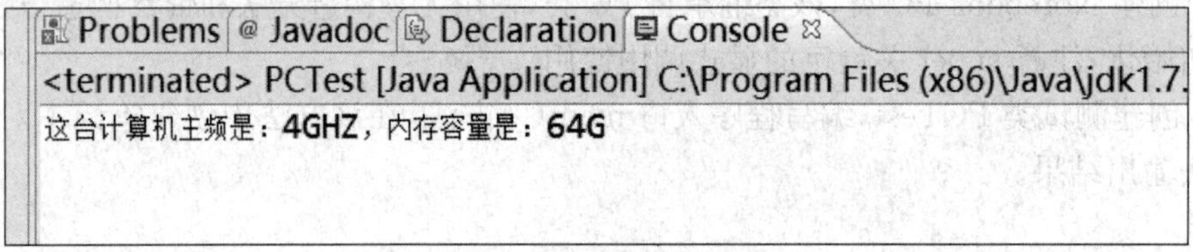

```
Problems  @ Javadoc  Declaration  Console
<terminated> PCTest [Java Application] C:\Program Files (x86)\Java\jdk1.7.
这台计算机主频是：4GHZ，内存容量是：64G
```

图 2.6　程序运行结果图

## 2.5　实验习题

1. 选择题

(1) 以下关于继承的叙述正确的是(　　)。
A. 在 Java 中类只允许单一继承
B. 在 Java 中一个类只能实现一个接口
C. 在 Java 中一个类不能同时继承一个类和实现一个接口
D. 在 Java 中接口只允许单一继承

(2) 有继承关系时用到的关键字是(　　)。
A. extend　　　　　B. extends　　　　　C. implements　　　　　D. implement

(3) Java 变量中，以下不属于复合类型的数据类型是(　　)。
A. 类　　　　　　　B. 字符型　　　　　　C. 数组型　　　　　　　D. 接口

(4) 若需要定义一个只在本类中使用的成员域或成员方法，应使用(　　)修饰符。
A. static　　　　　B. package　　　　　C. private　　　　　　D. public

(5) 对封装的理解正确的是(　　)。
A. 封装就是把对象的属性和行为结合成一个独立的单位。
B. 封装就是把对象完全隐蔽起来，不让外界访问。
C. 封装性是一个使用问题。
D. 封装和抽象是一回事。

(6) 在调用构造函数时(　　)。
A. 子类可以不加定义就使用父类的所有构造函数

B. 不管类中是否定义了何种构造函数,创建对象时都可以使用默认构造函数

C. 先调用父类的构造函数

D. 先调用形参多的构造函数

(7) 用于声明一个常量的修饰符是(　　)。

A. static　　　　B. abstract　　　　C. public　　　　D. final

(8) 在 Java 程序中定义一个类,类中有一个没有访问权限修饰的方法,则此方法(　　)。

A. 访问权限默认为 private　　　　B. 访问权限默认为 public

C. 访问权限默认为 protected　　　D. 都不是

2. 编程题

(1) 按以下要求编写程序

① 创建一个矩形类 Rectangle 类,包含宽 width 和高 height 两个成员变量。

② 在 Rectangle 中添加两个成员方法 calcGirth() 和 calcArea() 分别计算矩形的周长和面积。

③ 在 Rectangle 中创建一个无参构造方法和一个带参构造方法来完成矩形对象属性的初始化工作。

④ 编写测试类,生成若干个矩形并输出它们的周长和面积。

(2) 设计一个 Dog 类,该类具有名字、犬种、颜色三个属性;定义无参构造方法和带参构造方法来初始化类的这些属性;定义两个成员方法,一个用于输出 Dog 信息,一个用于输出狗的叫声。编写测试类使用 Dog 类创建多个 Dog 对象,并输出这些对象的信息和叫声。

(3) 按以下要求编写程序

① 定义父类汽车类 Mobile,具有名称属性 name,颜色属性 color;定义构造方法来对这些属性进行初始化;定义成员方法:行驶 run() 和刹车 stop() 方法。

② 定义两个子类,轿车类 Car 和卡车类 Truck,他们继承自 Mobile 类,定义各自的构造方法,通过调用父类构造方法实现该类属性的初始化。

③ 在轿车类 Car 和卡车类 Truck 中增加 addOil() 的方法,由于汽车和卡车所加油不同,所以 addOil() 方法的实现并不同,汽车添加汽油而卡车添加柴油。

④ 编写测试类,生成 Car 和 Truck 的对象并输出不同的方法输出对象的不同信息。

**参考文献:**

[1] 梁勇著 戴开宇译. JAVA 语言程序设计. 机械工业出版社. 2015,07.

[2] 边金良 孙红云编著. JAVA 程序设计教程和上机实验. 人民邮电出版社. 2015,05.

# 实验 3　类的多态性

## 3.1　知识点回顾

1. 多态性

多态性(polymorphism)一词来源于拉丁语 poly(表示多的意思)和 morphos(意为形态)，其字面的含义是多种形态。"多态性"原用于生物学，指在同一生物群体中，经常存在两种或者多种不连续的变异型或等位基因，也称之为遗传多态性。这一概念也被应用到面向对象的程序设计语言(如 Java)中来，指同一方法被不同的实例调用，将产生不同的运行结果。即不同的对象，收到同一消息可以产生不同的结果，多态性是面向对象程序设计的重要特征之一。面向对象的程序设计中利用多态性可以提高可扩充性和可维护性，大大提高程序的可复用性。

Java 通过方法的重写、重载和动态连接来实现多态性的。方法的重载实现的多态性又称为静态多态性，是在编译过程中就实现的。方法的重写实现的多态性则称为动态多态性，是在程序运行时才识别的。

2. 方法重载

方法重载(overloading)是让类以统一的方式处理不同类型数据的一种手段。Java 的方法重载，即在类中可以创建多个方法，它们具有相同的名字，但具有不同的参数，调用方法时通过传递给它们的不同个数和类型的参数来决定具体使用哪个方法。编译器根据不同的参数列表，对同名方法的名称做修饰，所以对于编译器来说这些同名方法是不同的。例如两个同名方法：int fun(int A)和 int funA(double B)，它们的地址在编译期间就已经确定了。就此而言，重载并不是真正的多态。重载的方法只能有相同的方法名，不能有相同的形参表，方法重载在类的构造函数中应用较多。需要注意的是，无法设计出参数个数和参数类型完全相同，而只有返回类型不同的重载。

3. 方法重写

方法重写(overriding)是指在子类中定义的一个方法，其名称、返回类型及参数列表与父类中某个方法的名称、返回类型及参数列表一致，就认为子类的方法重写了父类的方法。方法的重写是子类和父类之间的关系，重写的方法有相同的方法名和形参表，重写时区分方法的是根据调用他的对象。虽然子类能够通过重写把父类的状态和行为改变成自己需要的状态和行为，但是仍可通过关键字 super 调用父类中的方法及成员变量。

4. 方法的重写与重载的区别

(1) 方法的重写是子类和父类之间的关系，而重载是同一类内部多个方法间的关系；

(2) 方法的重写一般是两个方法间的，而重载时可能有多个重载方法；

(3) 重写的方法有相同的方法名和形参表，而重载的方法只能有相同的方法名，不能有相同的形参表；

(4) 重写时区分方法的是根据调用他的对象,而重载是根据形参来决定调用的是哪个方法;

(5) 用 final 修饰的方法是不能被子类重写的,只能被重载。

5. Java 语言中多态实现的机制

Java 中多态含义更多的是指用来在运行时动态选择调用的方法或对象的机制,正是方法的重写使得多态得以实现。Java 中运用方法的重写实现多态主要有以下两种机制:

(1) 通过父类或抽象类对象动态调用子类的方法

假如一个类有很多子类,并且这些子类都重写了父类的某个方法来分别产生不同的行为,如何能使一个父类对象动态调用属于子类的该方法呢?这就涉及对象的转型问题。允许把子类创建的一个对象的引用放到一个父类对象中,此时称该父类对象是子类对象的上转型对象(实际是将子类对象的指针赋值给父类对象的指针)。通过上转型对象调用属于子类的方法时就可能具有多种形态。由于对象是在运行时由其所属的类动态生成的,它所调用的方法的地址也是在运行期动态绑定的。

下面以一个子类对象与父类对象(上转型对象)的具体实例来阐述多态的应用。例:设计一个企业用户上网收费程序,User 类表示所有用户的基类,charge()方法用于输出上网应收金额,但具体金额根据具体客户类型确定。目前客户分为计时用户和包年用户,分别用 User 类的两个子类表示。在类 PolyTest 中,通过上转型对象调用子类方法实现对不同类别用户的费用计算。

```java
class User{
    private String userName; //用户名
    User(String name){this.userName = name;}
    String getName(){return userName;}
    void charge(){}; //收费金额计算由子类确定
}
class MinuteUser extends User{
    double unitPrice;
    int minutes;
    MinuteUser(String name,double price,int minutes)
    {
        super(name); //调用父类构造方法
        unitPrice = price;this.minutes = minutes;
    }
    void charge(){ //重写父类方法
        System.out.println(getName() + " is:" + unitPrice * minutes);
    }
}
class YearUser extends User{ //包年用户
    double mCharge;
    YearUser(String name,double mcharge){
```

```java
        super(name);
        this.mCharge = mcharge;
    }
    void charge(){
        System.out.println(getName() + " is:" + this.mCharge);
    }
}
public class PolyTest{
    public static void main(String[] args){
        User mu = new MinuteUser("chen",0.05,315);
        User yu = new YearUser("wang",360);
        mu.charge();
        yu.charge();
    }
}
```

上例中,通过动态绑定,charge()方法根据实际用户类型输出相应的金额。若后续还需增加新的用户类型,只需新定义一个继承 User 类的子类,并根据实际需求重写其中的 charge() 方法即可。在上述例子中可以看到,上转型对象的实体(如 mu)是由子类负责创建的,但由父类引用,因此子类的一些属性和功能对于上转型对象来说是被限制的。关于上转型对象有如下结论:

(1)上转型对象不能操作子类新增的成员变量,不能调用子类新增的方法。

(2)上转型对象调用重写的成员变量时,该变量总是反映父类的性质。

(3)上转型对象调用重写的静态成员方法时,仍是反映父类的性质,并不具有多态性。

(4)只有子类重写类父类的非静态方法时,上转型对象调用这个方法时调用的是子类重写方法,体现类多态性。

(5)若将上转型对象再强制转换为相应子类的对象,它又具备了子类所有的属性和方法。

(6)上转型对象的方法可以在沿着继承链的多个子类中实现,具体调用时 JVM 会沿着继承链依次查找方法的具体实现,一旦找到就停止查找并调用该方法。方法调用优先级由高到低依次为:this.show(O)、super.show(O)、this.show((super)O)、super.show((super)O)。

在很多应用中,类层次的顶层类并不具备下层类的一些具体实现功能。我们可在超类中将这些方法声明为没有实现的抽象方法,含有抽象方法的类就叫抽象类。这样,就可通过顶层类提供统一处理该类层次的方法。例如:设计一个方法求不同形状的物体的面积。首先设计一个 Shape 类,并定义一个抽象方法 showarea(),因为 Shape 类它不代表某些具体的对象。

```java
abstract class Shape    //抽象类
{
    private double x,y;
    public Shape(double x1,double y1){x = x1;y = y1;}
    abstract public void showarea();
}
```

接着,为不同形状的物体定义子类并继承 shape 类。

```
class Rectangle extends shape{
    private double w,h;
    public Rectangle(double x,double y,double w1,double h1)
    { super(x,y);w = w1;h = h1;}
    public void showarea()
    { System.out.println("Area of rectangle is " + w * h);}
}
Public class Test
{   public static void disparea(Shape r){r.showarea();} //多态程序段
    public static void main(String args[])
    { Rectangle r = new Rectangle(1,1,2,3);disparea(r);}
}
```

在上面例子中,可按如下的方式统一地处理该类层次的计算并显示面积的功能。

一般地,在顶层的超类中,总有很多的这种抽象方法,它为其他子孙类用抽象机制实现多态性提供了一个统一的界面。这是设计多态性经常用到的模式。

由于抽象类(如 Shape)不能产生对象,因此方法 disparea( )用来统一操作派生类 ReCtangle 和 Circle 的对象。类 Shape 的设计尽管是用继承性语法表达的,但它的主要目的不是为代码共享而设计的,而是为使用多态性而设计的,它是另一个维度的抽象。

(2) 接口

接口是 Java 中支持多态性的另一重要概念。接口允许多继承,也就是说,当类或接口用关键字 implements 或 extends 从接口继承时,它可以同时有多个父接口,这一点与类继承是不同的。通过接口继承我们可以实现接口的组合与扩充。接口常常被用来为具有相似功能的一组类,对外提供一致的服务接口,这一组类可以是相关的,也可以是不相关的,而抽象类则是为一组相关的类提供一致的服务接口。所以,接口往往比抽象类具有更大的灵活性。

我们在使用抽象类和接口时,必须注意以下 3 个方面:

(1) 抽象类中可以包含方法的声明,也可以提供方法的实现代码,而接口中只能提供方法声明,不可以有任何实现代码;

(2) 抽象类与其子类之间存在层次关系,而接口与实现它的类之间则不存在任何层次关系;

(3) 抽象类只能被单继承,而接口可以被多继承。

Java 的接口既和继承有关又和多态相关,而且是众多的面向对象程序设计语言不支持的,是 Java 的一个重要特征。

在上面的例子中,将 Shape 定义为抽象类是恰当的。但当作为一个类型声明的时候,接口拥有抽象类无法与之较量的灵活性。我们来看一个具体例子,下面是接口 Action 的声明:

```
interface Action
{   void close( );
    void open( );
```

```
}
class Door implements Action
{   public void close( ) { System.out.println(" Door close! ");}
    public void open( ) { System.out.println(" Door open! ");}
}
class Lamp implements Action
{   public void close( ) { System.out.println(" Lamp close! ");}
    public void open( ) { System.out.println(" Lamp open! ");}
}
public class Test
{   public static void close (Action r) { r.close();}   //多态程序段
    public static void main(String args[])
    {   Door r1 = new Door ();close(r1);
    Lamp r2 = new Lamp();close(r2);
     }
}
```

从上述代码可以看出，接口可以派生不相关的两个类，即门类 Door 和灯 Lamp，但它们实现相同行为 open 或 close。因此接口可以实现更加灵活多样的多态性。接口还可以避免盲目类继承所带来的潜在危险。比如，当我们不了解一个类的全部属性和方法，对类进行盲目扩充时，同样可能产生冗余代码，而且在继承我们所需的属性和方法的同时，有可能意外继承一些不需要的方法和属性，这在编译中不会报错，但方法被类对象调用时会产生一些不可控制的后果。而接口可以避免这些问题的产生，它要求实现接口的所有类必须实现接口中的所有方法，否则该类不可以被实例化。可以说接口是接口用户（使用接口的程序）和接口实现者（实现接口的类）之间一致的约定，比抽象类更加安全，清淅，不存在盲目性。就这些原因来说，当不需要为一组类提供公用实现代码时，我们优先考虑接口，以使用 Java 语言的多态性。抽象类和接口是 Java 语言中的二个重要的对象引用类型，是 Java 程序设计使用多态性的基础。

## 3.2　实验练习

### 3.2.1　实验任务 1

● 实验任务

本项任务是实现整数、字符串及小数 3 种数据类型的求和运算。

● 实验要点

（1）理解为什么需要方法重载。

（2）能够正确使用方法重载编程。

● 实验分析

通过任务描述可以看出实现本任务需要使用方法重载技术，设计 3 个名称相同但输入数据和返回值类型不同的方法,根据输入数据类型的不同调用相应的方法。

● 实验主要代码

根据上面分析,定义一个类 NumberSum,代码如下:

```java
package overload;

public class NumberSum {
    public int Sum(int a, int b)
    {
        System.out.println(a + b);
        return a + b;
    }
    public String Sum(String a, String b)
    {
        System.out.println(a + "," + b);
        return a + b;
    }
    public double Sum(double a, double b)
    {
        System.out.println(a + b);
        return a + b;
    }
}
```

编写测试类 OverloadTest,代码如下:

```java
package overload;
public class OverloadTest {

    /**
     * @param args
     */
    public static void main(String[] args) {
        // TODO Auto-generated method stub
        NumberSum sumTest = new NumberSum();
        sumTest.Sum(4,5);
        sumTest.Sum("hello","how are you!");
        sumTest.Sum(4.5,5.7);
    }
}
```

OverloadTest.java 文件运行结果如图 3.1 所示。

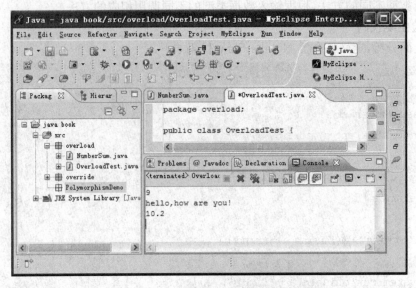

图 3.1　OverloadTest.java 文件运行结果

### 3.2.2　实验任务 2

● 实验任务

本项任务是实现下面对人类社会不同角色的描述。

（1）雇员是人类社会中的一种角色，每个雇员信息描述中应包括名字和工作报酬。

（2）学生也是人类社会中的一种角色，每个学生描述信息中应包括名字和就读的学校名。

● 实验要点

（1）理解为什么需要方法重写。

（2）理解方法重写与继承的关系。

（3）能够正确使用抽象类实现方法重写。

● 实验分析

通过任务描述可以看出需要使用方法重写技术：

（1）人类社会中的不同角色都有一个共同点，即都有名字。因此，可以把获取名字作为父类的抽象方法，每个子类继承该方法。

（2）不同角色都有自己的信息描述，因此子类需要重写父类的信息描述方法。

● 实验主要代码

根据上面分析，定义一个抽象类 People，代码如下：

```
package override;

abstract class People{
    private String name;
    public People(String n)
    {
```

```
        name = n;
    }
    public String getName()
    {
        return name;
    }
    public abstract String getDescription();
}
```

定义子类 Employee，代码如下：

```
package override;

class Employee extends People{
    private double salary;
    public Employee(String n, double s)
    {
        super(n);
        salary = s;
    }
    public double getSalary()
    {
        return salary;
    }
    public void raiseSalary(double byPercent)
    {
        double raise = salary * byPercent /100;
        salary + = raise;   }
    public String getDescription()
    {
        return String.format("'s salary is $ %.2f", salary);
    }
}
```

定义子类 Student，代码如下：

```
package override;

class Student extends People
{
    /**
        @param n the student's name
        @param m the student's school
```

```java
     */
    private String school;
    public Student(String n, String m)
    {
        // pass n to superclass constructor
        super(n);
        school = m;
    }
    public String getDescription()
    {
        return "'s school is " + school;
    }
}
```

编写测试类 PeopleTest,代码如下:

```java
package override;

public class peopleTest {
    /**
     * @param args
     */
    public static void main(String[] args) {
        People[] people = new People[2];
        // fill the people array with Student and Employee objects
        people[0] = new Employee("Harry Hacker", 5000);
        people[1] = new Student("Maria Morris", "University of Oxford");
        // print out names and descriptions of all Person objects
        for (People p : people)
            System.out.println(p.getName() + p.getDescription());
    }
}
```

PeopleTest.java 程序运行结果如图 3.2 所示。

实验3 类的多态性

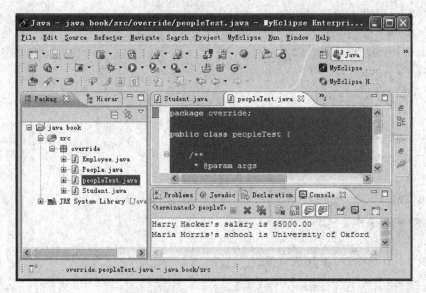

图 3.2　PeopleTest.java 程序运行结果

## 3.3　项目实战 1

### 3.3.1　项目描述

请用 Java 的面向对象技术实现如下描述：狗、熊和猴都是动物，但狗、熊和猴都有 4 条腿。杂技团训练动物，不同的动物训练内容不一样：狗倒立、数数，熊打拳，猴子骑车子、鞠躬和翻跟头。

### 3.3.2　项目分析

通过项目描述可以看出：
(1) 狗、猴和熊都是动物，可以定义一个 Animal 父类，再分别定义 Dog、Monkey 和 Bear3 个子类。
(2) 定义 Train 类，根据对象的训练内容数目的不同执行不同的训练方法。

### 3.3.3　项目编写

根据上面分析，定义一个抽象类 Animal，代码如下：

```
package PolymorphismDemo;

class Animal {
```

```java
    private int numLegs;
    private String type;
    public Animal(int n,String t){
        numLegs = n;
        type = t;
    }
    public int getNumLegs(){
        return numLegs;
    }
    public String getType(){
        return type;
    }
    public void getDescription(){
        System.out.println("动物世界!");
    }
}
```

再定义 3 个子类分别是:Dog 类,代码如下:

```java
package PolymorphismDemo;

public class Dog extends Animal{
    public Dog(int n,String t){
        super(n,t);
    }
    public void getDescription()
    {
        System.out.println(super.getType() + ",有" + super.getNumLegs() + "条腿!");
    }
}
```

Bear 类,代码如下:

```java
package PolymorphismDemo;

public class Bear extends Animal{
    public Bear(int n,String t){
        super(n,t);
    }
    public void getDescription()
    {
        System.out.println(super.getType() + ",有" + super.getNumLegs() + "条腿!");
```

## 实验 3 类的多态性

```
        }
}
```

Monkey 类,代码如下:

```
package PolymorphismDemo;

public class Monkey extends Animal{
    public Monkey(int n,String t){
        super(n,t);
    }
    public void getDescription()
    {
        System.out.println(super.getType() + ",有" + super.getNumLegs() + "条腿!");
    }
}
```

接着定义一个训练方法 Training 类,代码如下:

```
package PolymorphismDemo;

public class Training {
    public Training(){
        System.out.println("什么都不做!");
    }
    public void TrainingContent(Animal animal,String s1){
        System.out.println(animal.getType() + "的训练内容是" + s1);
    }
    public void TrainingContent(Animal animal,String s1,String s2){
        System.out.println(animal.getType() + "的训练内容是" + s1 + "和" + s2);
    }
    public void TrainingContent(Animal animal,String s1,String s2,String s3){
        System.out.println(animal.getType() + "的训练内容是" + s1 + "、" + s2 + "和" + s3);
    }
}
```

编写测试类 PloymorphismTest 类,代码如下:

```
package PolymorphismDemo;

public class PolymorphismTest {
        public static void main(String[] args) {
```

```java
// TODO Auto-generated method stub
Animal[] animal = new Animal[3];
animal[0] = new Dog(4,"狗");
animal[1] = new Bear(4,"熊");
animal[2] = new Monkey(4,"猴");
for(Animal a :animal){
    a.getDescription();
}
Training train = new Training();
train.TrainingContent(animal[0],"倒立","数数");
train.TrainingContent(animal[1],"打拳");
train.TrainingContent(animal[2],"骑车","鞠躬","翻跟头");
}
}
```

最后,测试类 PloymorphismTest 类运行结果如图 3.3 所示。

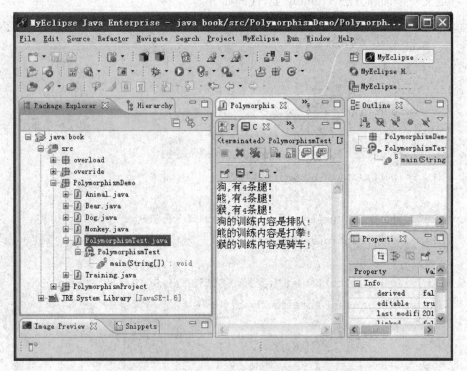

图 3.3　测试类 PloymorphismTest 类运行结果

## 3.4 项目实战2

### 3.4.1 项目描述

请用Java的面向对象技术实现如下描述：计算机的主机包含主板、网卡和声卡等部件，它们都有各自的厂商。其中声卡和网卡都通过标准接口与主板相连，他们的工作都是由主板进行控制。

### 3.4.2 项目分析

通过项目描述可以看出：
（1）可以定义一个PCI接口，该接口包含开始和结束两个操作。
（2）分别定义继承了PCI接口的SoundCard和NetCard两个类，在这两个类中需要实现开始和结束两个具体操作行为，同时还需包含显示生产厂家的构造方法。
（3）定义MainBoard类，该类负责实现调用基于PCI接口的其他类的开始和结束操作。

### 3.4.3 项目编写

根据上面分析，定义一个接口InterfacePCI，代码如下：

```java
package PCI;
public class MainBoard {
    /**
     * 主板类,放置PCI插槽
     */
    private String productor;//生产商
    //封装productor
    public String getProductor()
    {
        return productor;
    }
    //构造方法
    public MainBoard(){
        this.productor = "不详";
    }
    public MainBoard(String productor){
        this.productor = productor;
```

```java
    }
    //实现 PCI 接口的调用
    public void usePCICard(Interface PCI)
    {
        PCI.start();
        PCI.stop();
    }
}
```

接着,分别定义 NetWorkCard 和 SoundCard 两个类,代码如下:

NetWorkCard 类:
```java
package PCI;

public class NetWorkCard   implements InterfacePCI {
    /**
     * 网卡类,实现了 InterfacePCI 接口
     */
    private String productor;//生产商
    public String getProductor()
    {
        return productor;
    }
    //构造方法
    public NetWorkCard(){
        this.productor = "不详";
    }
    public NetWorkCard(String productor)
    {
        this.productor = productor;
    }
    //实现 InterfacePCI 接口中的 start()方法
    public void start()
    {
        System.out.println("正在发送数据...");
    }
    //实现 InterfacePCI 接口中的 stop()方法
    public void stop()
    {
        System.out.println("网卡已工作结束");
    }
}
```

SoundCard 类:
```java
package PCI;
public class SoundCard implements InterfacePCI{
    /**
     * 声卡类,实现接口 InterfacePCI
     */
    //构造函数
    public SoundCard(){
        this.productor = "不详";
    }
    public SoundCard(String productor){
        this.productor = productor;
    }
    private String productor;//生产商
    public String getProductor()
    {
        return productor;
    }

    //实现 start()方法
    public void start()
    {
        System.out.println("嘀嘀嘀...");
    }
    //实现 stop()方法
    public void stop()
    {
        System.out.println("声卡已工作结束");
    }
}
```

再定义 MainBoard 类,代码如下:
```java
package PCI;

public class MainBoard {
    /**
     * 主板类,放置 PCI 插槽
     */
    private String productor;//生产商
    //封装 productor
```

```java
    public String getProductor()
    {
        return productor;
    }
    
    //构造方法
    public MainBoard(){
        this.productor = "不详";
    }
    public MainBoard(String productor){
        this.productor = productor;
    }
    
    //实现 PCI 接口的调用
    public void usePCICard(InterfacePCI PCI)
    {
        PCI.start();
        PCI.stop();
    }
}
```

编写测试类 TestPCI 类,代码如下:

```java
package PCI;

public class TestPCI {
    /**
     * @param args
     */
    public static void main(String[] args) {
        // TODO Auto-generated method stub
        
        //创建一块声卡
        SoundCard sound = new SoundCard("Realtek");
        
        //创建一块网卡
        NetWorkCard net = new NetWorkCard("DLink");
        
        //创建一个主板
        MainBoard board = new MainBoard();
        System.out.println("生产厂家是" + board.getProductor() + "的");
```

```
      //主板调用声卡
      System.out.println("主板开始调用" + sound.getProductor() + "生产的声卡");
      board.usePCICard(sound);
      //主板调用网卡
      System.out.println("主板开始调用" + net.getProductor() + "生产的网卡");
      board.usePCICard(net);
   }
}
```

最后，测试类 TestPCI 类运行结果如图 3.4 所示。

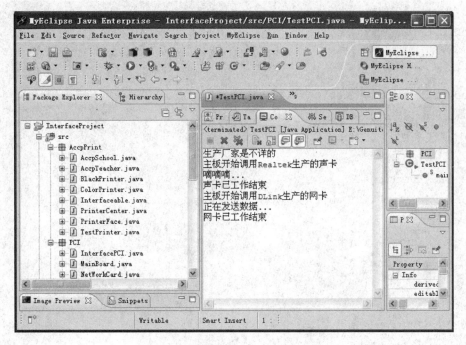

图 3.4　测试类 TestPCI 类运行结果

## 3.5　实验习题

1．选择题

（1）阅读下面代码：

```
public class Parent{
   public int addValue( int a, int b){
      int s;
      s = a + b;
      return s;
   }
}
```

```
}
class Child extends Parent{
}
```

下面的哪些方法可以加入类 Child 中？（　　）
  A. int addValue(int a,int b){//做某事情…}
  B. public void addValue( ){//做某事情…}
  C. public int addValue(int a){//做某事情…}
  D. public int addValue(int a,int b)throws Exception{//做某事情…}

2. 填空题

```
public class A {
    public String show(D obj){
        return ("A and D");
    }
    public String show(A obj){
        return ("A and A");
    }
}
public class B extends A{
    public String show(B obj){
        return ("B and B");
    }public String show(A obj){
        return ("B and A");
    }
}
public class C extends B{
    public String show(C obj){
        return ("C and C");
    }
    public String show(B obj){
        return ("C and B");
    }
}
public class D extends B{
    public String show(D obj){
        return ("D and D");
    }
    public String show(B obj){
        return ("D and B");
    }
```

```
}
public class polyTest {
    public static void main(String[] args) {
        // TODO Auto-generated method stub
        A a1 = new A();
        A a2 = new B();
        B b = new B();
        C c = new C();
        D d = new D();
        System.out.println(a1.show(b));     ①
        System.out.println(a1.show(c));     ②
        System.out.println(a1.show(d));     ③
        System.out.println(a2.show(b));     ④
        System.out.println(a2.show(c));     ⑤
        System.out.println(a2.show(d));     ⑥
        System.out.println(b.show(b));      ⑦
        System.out.println(b.show(c));      ⑧
        System.out.println(b.show(d));      ⑨
    }
}
    输出结果是:①        ②          ③          ④          ⑤
              ⑥        ⑦          ⑧          ⑨
```

3. 编程题

编写 Shape 类、Rectangle 类和 Circle 类。其中,Shape 类是父类,其他两个类是子类。Shape 类包含 2 个属性:x 和 y,以及一个方法 draw();Rectangle 类增加了两个属性:长度和宽度;Circle 类增加了一个属性:半径。使用一个主方法来测试 Shape 中的数据和方法可以被子类继承。然后分别在两个子类中重写 draw()方法并实现多态。

4. 代码分析题

当你尝试编译并运行如下代码,将会输出什么结果?并解释。本题代码如下:

```
class SupClass {
    int i = 1;
    void go()
    {
        System.out.println("SupClass + go()");
    }
}
class SubClass extends SupClass {
    int i = 2;
    void go()
```

```java
    {
        System.out.println("SubClass + go()");
    }
    void go(int i)
    {
        System.out.println("SubClass + go(int i)");
    }
}
public class MainClass {

    public static void main(String[] args) {
        // TODO Auto-generated method stub
        SupClass a = new SubClass();
        System.out.println(a.i);
        a.go();
        a.go(1);
        a = (SubClass)a;
        System.out.println(a.i);
        a.go();
        a.go(2);
        SubClass b = (SubClass)a;
        System.out.println(b.i);
        b.go();
        b.go(2);
    }
}
```

## 3.6 小结

多态的优点:

(1) 可替换性(substitutability)。多态对已存在代码具有可替换性。例如,多态对圆 Circle 类工作,对其他任何圆形几何体,如圆环,也同样工作。

(2) 可扩充性(extensibility)。多态对代码具有可扩充性。增加新的子类不影响已存在类的多态性、继承性,以及其他特性的运行和操作。实际上新加子类更容易获得多态功能。例如,在实现了圆锥、半圆锥以及半球体的多态基础上,很容易增添球体类的多态性。

(3) 接口性(interface-ability)。多态是超类通过方法签名,向子类提供了一个共同接口,由子类来完善或者覆盖它而实现的。如上述项目分析 2 所示。接口 InterfacePCI 规定了两个实现多态的接口方法,start()以及 stop()。子类 NetWorkCard 和 SoundCard 为了实现多态,完善或者覆盖这两个接口方法。

(4) 灵活性(flexibility)。它在应用中体现了灵活多样的操作,提高了使用效率。

(5) 简化性(simplicity)。多态简化对应用软件的代码编写和修改过程,尤其在处理大量对象的运算和操作时,这个特点尤为突出和重要。

建议和说明:

(1) Java 中除了 static 和 final 方法外,其他所有的方法都是运行时绑定的。private 方法都被隐式指定为 final 的,因此 final 的方法不会在运行时绑定。当在派生类中重写基类中 static、final 或 private 方法时,实质上是创建了一个新的方法。

(2) 在派生类中,对于基类中的 private 方法,最好采用不同的名字。

(3) 包含抽象方法的类叫作抽象类。注意定义里面包含这样的意思,只要类中包含一个抽象方法,该类就是抽象类。抽象类在派生中就是作为基类的角色,为不同的子类提供通用的接口。

(4) 对象清理的顺序和创建的顺序相反,当然前提是自己想手动清理对象,因为大家都知道 Java 垃圾回收器。

(5) 在基类的构造方法中小心调用基类中被重写的方法,这里涉及对象初始化顺序。

(6) 构造方法是被隐式声明为 static 方法。

(7) 多态在目前很多开源框架都有实际运用,如 Spring 开源框架结合配置文件的使用,利用反射,动态的调用类,同时不用修改源代码,直接添加新类和修改配置文件,不需要重启服务器便可以扩展程序。

**参考文献:**

蓝雯飞、周俊陈、淑清.JAVA 语言的多态性及其应用研究.计算机系统应用 ,2005(4).

# 实验 4 接 口

## 4.1 知识点回顾

1. 抽象类

抽象类是使用修饰符 abstract 修饰的类,抽象方法是使用修饰符 abstract 修饰的方法。在类的前面添加"abstract"关键字,可以创建抽象类。在方法的前面添加"abstract"关键字,可以创建抽象方法。示例如下:

```
//抽象类
public abstract class Animal
{
    //抽象方法
    public abstract void makeSound();
}
```

抽象类和抽象方法的一些特性:

(1) 抽象类不能用来创建类实例,只能用于派生出子类。而子类必须实现抽象类中的所有抽象方法,否则,子类仍然是抽象类。

(2) 抽象方法是不能有实现代码。

(3) 同一个抽象类中派生出来的几个子类可以包含相同的办法,这些方法的实现代码可以有所不同。抽象方法的修饰权限只能被定义为 public 和 protect,不能被定义为 private。例如:

```
public class Dog extends Animal
{
    public void makeSound()
    {
        System.out.println("汪!汪!汪!");
    }
}
public class Cat extends Animal
{
    public void makeSound ()
    {
        System.out.println("喵!喵!喵!");
```

(4) 抽象类中可以存在抽象方法,也可以存在非抽象方法,但是存在抽象方法的类一定是抽象类。例如:

```java
public abstract class Animal
{
    //抽象方法
    public abstract void makeSound();
    //非抽象方法
    public void makeRun()
    {
        System.out.println("可以奔跑");
    }
}
public class Dog extends Animal
{
    public void makeSound()
    {
        System.out.println("汪!汪!汪!");
    }
}
```

2. 接口

接口是对抽象类的进一步扩展,接口是抽象类的变体,接口中的方法都是未实现的,目的是在实现接口的类之间建立一种协议。接口中的所有方法都是抽象的,没有一个有程序体。接口只可以定义 static final 成员变量。

接口的好处是,它给出了屈从于 Java 技术单继承规则的假象。当类定义只能扩展出单个类时,它能实现所需的多个接口。

接口的实现与子类相似,当类实现接口时,它定义所有这种接口的方法。然后,它可以在实现了该接口的类的任何对象上调用接口的方法。由于有抽象类,它允许使用接口名作为引用变量的类型。通常的动态联编将生效。引用可以转换到接口类型或从接口类型转换,instanceof 运算符可以用来决定某对象的类是否实现了接口。

接口是通过"interface"关键字来进行定义的。例如:

```java
public interface Swim
{
    public static final int id = 1;
    public static final int age = 28;
    void swim();
}
```

接口的一些特性:

(1) 接口中的方法修饰符默认是 public 和 abstract。

（2）接口中的方法修饰符不能为 native、private、static、final、synchronized 等。

（3）接口中的属性成员修饰符默认是 public、static、final。

（4）接口可以有父接口，Java 中不存在类的多继承，但是允许接口的多继承。例如：

```java
interface CanFight
{
    void fight();
}
interface CanSwim
{
    void swim();
}
interface CanFly
{
    void fly();
}
class Hero implements CanFight,CanSwim,CanFly
{
    public void fight()
    {
        System.out.println("擅长打战");
    }
    public void swim()
    {
        System.out.println("擅长游泳");
    }
    public void fly()
    {
        System.out.println("擅长飞行");
    }
}
```

## 4.2 实验练习

### 4.2.1 实验任务1

● 实验任务

假设一所高校里有不同身份的人，他们有不同的行为，其对应规则如下：

| 身份 | 行为 |
|---|---|
| 教师 | 讲课 |
| 学生 | 学习 |
| 实验员 | 管理实验室 |

要求编写一个接口 Person，并有行为 action，定义教师、学生、实验员三个类继承接口 Person，并实现各自的方法。

● 实验要点

（1）掌握接口的定义格式。
（2）掌握类对接口的继承。

● 实验分析

（1）首先要定义 Person 接口，接口中定义一个方法：action。
（2）Teacher 类继承 Person 接口，并重写 action 方法。
（3）Student 类继承 Person 接口，并重写 action 方法。
（4）Lab 类继承 Person 接口，并重写 action 方法。
（5）编写测试类 PersonTest，创建 Teacher 类实例、Student 类实例和 Lab 类实例，通过父接口 Person 对它们进行访问。

● 实验主要代码

根据上面分析，代码如下：

```
Interface Person{
    void action();
}

class Teacher implements Person{
    public void action(){
        System.out.println("教师在讲课!");
    }
}

class Student implements Person{
    public void action(){
        System.out.println("学生在学习!");
    }
}

class Lab implements Person{
    public void action(){
        System.out.println("实验员在管理实验室!");
    }
```

```
}
public class PersonTest{
    public static void main(String[] args)
    {
        Person p = new Teacher ();
        p.action();
        p = new Student();
        p.action();
        p = new Lab();
        p.action();
    }
}
```

运行结果如图 4.1 所示。

```
教师在讲课!
学生在学习!
实验员在管理实验室!
```

**图 4.1  测试类 PersonTest 运行结果**

现在如果要添加"教学秘书"和"辅导员"两类角色,只需添加"教学秘书"和"辅导员"类,而主类仍然可以拿来就用,无需进行更多的修改。此时就可以显示出接口的作用了。

在上面的程序中添加如下两个类即可。

```
class Secretary implements Person{
    public void action(){
        System.out.println("教学秘书在排课!");
    }
}

class Counsellor implements Person{
    public void action(){
        System.out.println("辅导员在和同学们谈心!");
    }
}
```

### 4.2.2  实验任务 2

● 实验任务

编写 Shape 接口、Rectangle 类和 Circle 类。其中,Shape 是父接口,其他两个类是子类。

Shape 有两个方法一个求周长,一个求面积;Rect 类增加了两个属性:长度和宽度;Circle 类增加了一个属性:半径。使用一个主方法来测试 Shape 中的数据和方法可以被子类继承,并实现类的多态性。

● 实验要点

(1)掌握接口的定义格式。

(2)掌握类对接口的继承。

(3)掌握类的多态性。

● 实验分析

(1)首先要定义 Shape 接口,接口中定义两个方法:Perimeter 和 Area。

(2)Rect 类继承 Shape 接口,并且定义两个私有属性:length 和 wide,并重写 Perimeter 和 Area 方法。

(3)Circle 类继承 Shape 接口,并且定义一个私有属性:radius,并重写 Perimeter 和 Area 方法。

(4)编写测试类 Test,创建 Rect 类实例和 Circle 类实例,通过父接口 Shape 对它们进行访问,实现了多态性。

● 实验主要代码

根据上面分析,代码如下:

```
interface Shape
{
    //周长
    void Perimeter();
    //面积
    void Area();
}
//长方形类
class Rect implements Shape
{
    private int length;
    private int wide;
    //重写 Area 方法
    public void Area()
    {
        System.out.println("长方形长为:" + this.length + "  宽为:" + this.wide + "它的面积是:" + (this.length * this.wide));
    }
    //重写 Perimeter 方法
    public void Perimeter()
    {
```

```java
            System.out.println("长方形长为:" + this.length + "  宽为:" + this.wide + " 它的周长是:" + ((this.length * 2) + (this.wide * 2)));
        }
        //获得长度
        public int getLength()
        {
            return length;
        }
        //设置长度
        public void setLength(int length)
        {
            this.length = length;
        }
        //获得宽度
        public int getWide()
        {
            return wide;
        }
        //设置宽度
        public void setWide(int wide)
        {
            this.wide = wide;
        }
        //构造方法初始化
        public Rect(int length, int wide)
        {
            setLength(length);
            setWide(wide);
        }
}
//圆类
class Circle implements Shape
{
        private double radius;
        //重写 Area 方法
        public void Area()
        {
            System.out.println("圆的半径为:" + this.radius + " 它的面积是:" + (3.14 * this.radius * this.radius));
```

```java
        }
        //重写Perimeter方法
        public void Perimeter()
        {
            System.out.println("圆的半径为:" + this.radius + " 它的周长是:" + (2 * 3.14 * this.radius));
        }
        //获得半径
        public double getRadius()
        {
            return radius;
        }
        //设置半径
        public void setRadius(double radius)
        {
            this.radius = radius;
        }
        //构造方法初始化
        public Circle(double radius)
        {
            setRadius(radius);
        }
}
//测试类
public class Test
{
    public static void main(String[] args)
    {
        Shape[] Shape = new Shape[2];
        Rect a = new Rect (5, 5);
        Circle b = new Circle(8);
        Shape[0] = a;
        Shape[1] = b;
        for(int i = 0; i<2; i++)
        {
            Shape[i].Perimeter();
            Shape[i].Area();
        }
```

    }
}

运行结果如图 4.2 所示。

```
长方形长为：5    宽为：5  它的周长是：20
长方形长为：5    宽为：5  它的面积是：25
圆的半径为：8.0  它的周长是：50.24
圆的半径为：8.0  它的面积是：200.96
```

图 4.2  测试类 Test 运行结果

### 4.2.3  实验任务 3

● 实验任务

通过接口继承，定义 Duck 类，能够实现鸭叫和飞行的行为，并通过测试类来进行测试。

● 实验要点

(1) 能够正确使用接口的继承。
(2) 理解方法重写与继承的关系。
(3) 能够把多态性和接口继承结合起来使用。

● 实验分析

(1) 首先要定义 QuackBehavior 接口，接口中定义一个方法：quack。
(2) 定义 Quack 类继承 QuackBehavior 接口，并重写 quack 方法。
(3) 再定义 FlyBehavior 接口，接口中定义一个方法：fly。
(4) 定义 FlyWithWings 类继承 FlyBehavior 接口，并重写 fly 方法。
(5) 定义抽象类 Duck，定义方法：performFly 和 performQuack。
(6) 最后编写测试类 DuckSimulator，定义 display 和 eat 方法，并在 main 方法中使用多态性创建 duckOne 类实例，输出结果。

● 实验主要代码

```java
// QuackBehavior 接口   实现鸭子呱呱叫
interface QuackBehavior
{
    public void quack();
}
//定义 Quack 类,实现鸭子呱呱叫
class Quack implements QuackBehavior
{
    //实现接口中的 Quack() 方法
    public void quack()
    {
        System.out.println("呱呱呱, 我是能叫的鸭子");
```

```java
    }
}
//定义 MuteQuack 类
class  MuteQuack implements QuackBehavior
{
    //实现 QuackBehavior 接口中的 quack()方法
    public void quack()
    {
        System.out.println("沉默,我不会想鸭子一样叫");
    }
}
//定义 FlyBehavior 接口
interface FlyBehavior
{
    public void fly();
}
//定义 FlyWithWings 类
class FlyWithWings implements FlyBehavior
{
    //实现接口中的 fly() 方法
    public void fly()
    {
        System.out.println("飞起来了,我会飞");
    }
}
//定义 FlyNoWay 类
class FlyNoWay implements FlyBehavior
{
    public void fly()
    {
        System.out.println("我不会飞,呜呜呜!!");
    }
}
//定义 抽象类 Duck 类
abstract class Duck
{
    //接口成员变量
    FlyBehavior flyBehavior ;
    //接口成员变量
```

```java
    QuackBehavior quackBehavior ;
    public Duck()
    {
    }
    //抽象方法
    public abstract void display();
    public void performFly()
    {
        //委托给行为类
        flyBehavior.fly();
    }
    public void performQuack()
    {
        //委托给行为类
        quackBehavior.quack();
    }
}
//定义 DuckSimulator 类
public class DuckSimulator extends Duck
{
    //构造函数,实现飞行和鸭叫行为
    public DuckSimulator ()
    {
        //使用多态
        flyBehavior = new FlyWithWings();
        quackBehavior = new Quack();
    }
    //显示消息
    public void display()
    {
        System.out.println("我是鸭鸭,我是鸭鸭");
    }
    //觅食
    public void eat()
    {
        System.out.println("我开始觅食了");
    }
    public static void main(String[] args)
    {
```

```java
        //使用多态
        Duck duckOne = new DuckSimulator ();
        duckOne.performFly();
        duckOne.performQuack();
        duckOne.display();
        if(duckOne instanceof DuckSimulator)
        {
            //强制类型转换,将父类型转换成子类型
            DuckSimulator mallardDuck = (DuckSimulator) duckOne;
            System.out.println("");
            mallardDuck.display();
            mallardDuck.eat();
        }
    }
}
```

运行结果如图 4.3 所示。

```
飞起来了,我会飞
呱呱呱, 我是能叫的鸭子
我是鸭鸭,我是鸭鸭

我是鸭鸭,我是鸭鸭
我开始觅食了
```

图 4.3　DuckSimulator 类运行结果

### 4.2.4　实验任务 4

● 实验任务

首先创建接口 Usb,Wifi,然后通过接口继承,创建一个电脑类、相机类和 MP3 类,相机通过 Usb 接入电脑工作,MP3 通过 wifi 接入电脑工作,电脑可以使用相机和 MP3。

● 实验要点

(1) 能够正确使用接口的继承。
(2) 理解方法重写与继承的关系。
(3) 理解多接口的使用。

● 实验分析

(1) 首先要定义 Usb 接口,接口中定义启动方法 start 和停止方法 stop。
(2) 然后定义 Wifi 接口,接口中定义插入方法 insert。
(3) 创建电脑类 Computer,定义方法 useUsb, useWifi。
(4) 创建相机类 Camera,继承接口 Usb,重写启动方法 start,停止方法 stop。
(5) 创建 MP3 类 MyMP3,继承接口 Usb 和 Wifi,重写启动方法 start,停止方法 stop 和插入方法 insert。

（6）最后编写测试类 TestUsb，在 main 方法中创建 Computer，Camera，MyMP3 类实例，输出结果。

- **实验主要代码**

```
interface Usb
{
 public void start();
 public void stop();
}


interface Wifi extends Usb
{
 public void insert();
}


class Computer
{
    public void useUsb(Usb usb)
    {
     usb.start();
     usb.stop();
    }
    public void useWifi(Wifi wifi)
    {
     wifi.start();
     wifi.stop();
     wifi.insert();
    }
}


class Camera implements Usb
{
 public void start()
 {
   System.out.println("相机正常接入电脑");
 }
 public void stop()
 {
   System.out.println("相机正常停止工作");
 }
```

```java
}
//实现多接口
class MyMP3 implements Usb,Wifi
{
  public void start()
  {
    System.out.println("Mp3 正常接入电脑");
  }
  public void stop()
  {
    System.out.println("MP3 正常停止工作");
  }
  public void insert()
  {
    System.out.println("MP3 通过 wifi 连接上电脑");
  }
}

public class TestUsb {
  public static void main(String[] args) {
    //创建一个电脑
    Computer computer = new Computer();
    //创建一台相机
    Camera camera1 = new Camera();
    //创建一台 mp3
    MyMP3 Mp3 = new MyMP3();
    System.out.println("下面通过电脑实现这个接口");
    computer.useUsb(camera1);
    computer.useWifi(Mp3);
  }
}
```

运行结果如图 4.4 所示。

```
下面通过电脑实现这个接口
相机正常接入电脑
相机正常停止工作
Mp3正常接入电脑
MP3正常停止工作
MP3通过wifi连接上电脑
```

**图 4.4  测试类 TestUsb 运行结果**

## 4.3 项目实战1

### 4.3.1 项目描述

现在我们要开发一个应用,模拟移动存储设备的读写,即计算机与 U 盘、MP3、移动硬盘等设备进行数据交换。要求计算机能同这三种设备进行数据交换,并且以后可能会有新的第三方的移动存储设备,所以计算机必须有扩展性,能与目前未知而以后可能会出现的存储设备进行数据交换。各个存储设备间读、写的实现方法不同,U 盘和移动硬盘只有 Read 和 Write 这两个方法,MP3Player 还有一个 PlayMusic 方法。

### 4.3.2 项目分析

(1) 首先定义接口 MobileStorage,包含两个方法 Read 和 Write。
(2) 定义 U 盘、MP3、移动硬盘三个存储设备类继承此接口,并重写 Read 和 Write 方法。
(3) 定义 Computer 类中,包含一个类型为 MobileStorage 的成员变量,通过构造方法进行初始化。Computer 中包含两个方法:ReadData 和 WriteData。
(4) 定义测试类,通过多态性实现不同移动设备的读写。

### 4.3.3 项目编写

首先编写 MobileStorage 接口:

```java
interface MobileStorage
{
    void Read();
    void Write();
}
```

接下来是三个移动存储设备的具体实现代码。
U 盘:

```java
class FlashDisk implements MobileStorage
{
    public void Read()
    {
        System.out.println("Reading from FlashDisk......");
        System.out.println("Read finished!");
    }
    public void Write()
    {
```

```
      System.out.println("Writing to FlashDisk......");
      System.out.println("Write finished!");
    }
}
```

MP3：

```
class MP3Player implements MobileStorage
{
    public void Read()
    {
      System.out.println("Reading from MP3Player......");
      System.out.println("Read finished!");
    }
    public void Write()
    {
      System.out.println("Writing to MP3Player......");
      System.out.println("Write finished!");
    }
    public void PlayMusic()
    {
      System.out.println("Music is playing......");
    }
}
```

移动硬盘：

```
class MobileHardDisk implements MobileStorage
{
    public void Read()
    {
      System.out.println("Reading from MobileHardDisk......");
      System.out.println("Read finished!");
    }
    public void Write()
    {
      System.out.println("Writing to MobileHardDisk......");
      System.out.println("Write finished!");
    }
}
```

可以看出，U盘、MP3、移动硬盘三个存储设备类都实现了MobileStorage接口，并重写了各自不同的Read和Write方法。下面，我们来写Computer：

```java
class Computer
{
    MobileStorage usbDrive;
    public  Computer(MobileStorage myusbDrive)
    {
        usbDrive = myusbDrive;
    }
    public void ReadData()
    {
        usbDrive.Read();
    }
    public void WriteData()
    {
        usbDrive.Write();
    }
}
```

其中的 usbDrive 就是可替换的移动存储设备。

最后我们来测试我们的"电脑"和"移动存储设备"是否工作正常。具体代码如下：

```java
public class ComputerExample
{
    public static void main(String args[])
    {
    FlashDisk flashDisk = new FlashDisk();
    MP3Player mp3Player = new MP3Player();
    MobileHardDisk mobileHardDisk = new MobileHardDisk();
    System.out.println("I inserted my MP3 Player into my computer and copy some music to it:");
    Computer computer1 = new Computer(mp3Player);
    computer1.WriteData();
    System.out.println("Well, I also want to copy a great movie to my computer from a mobile hard disk:");
    Computer computer2 = new Computer(mobileHardDisk);
    computer2.ReadData();
    computer2.WriteData();
    System.out.println("OK! I have to read some files from my flash disk and copy another file to it:");
    Computer computer3 = new Computer(flashDisk);
    computer3.ReadData();
    computer3.WriteData();
```

        }
}

最后，测试类运行结果如图 4.5 所示。

```
I inserted my MP3 Player into my computer and copy some music to it:
Writing to MP3Player ......
Write finished!
Well,I also want to copy a great movie to my computer from a mobile hard disk:
Reading from MobileHardDisk ......
Read finished!
Writing to MobileHardDisk ......
Write finished!
OK!I have to read some files from my flash disk and copy another file to it:
Reading from FlashDisk......
Read finished!
Writing to FlashDisk......
Write finished!
```

图 4.5  测试类 ComputerExample 运行结果

## 4.4 项目实战 2

### 4.4.1 项目描述

我们要构建一个接口 IQueue，方法 put( )和 get( )为队列定义了接口，而没有定义实现细节。然后构建一个类 Queue 继承接口 IQueue，该类实现了一个简单的固定大小的字符队列。最后我们再构建一个循环队列 CircularQueue，在循环队列中当遇到内部的数组末尾时，get 和 put 索引将会自动返回到起点。这样，任何数目的元素都能够存储在一个循环队列中。

### 4.4.2 项目分析和编写

（1）创建一个名为 IQueue.java 的文件并把下面的接口定义放进该文件：

```java
public interface IQueue
{
    void put (char ch);
    char get ();
}
```

这个接口由两个方法：put( ) 和 get( )组成。
（2）创建一个名为 QueueDemo.java 的文件。
（3）通过添加 Queue 类创建 QueueDemo.java：

```java
class Queue implements IQueue
{
```

```java
    private char q[];
    private int putloc, getloc;
    public Queue(int size)
    {
        q = new char[size + 1];
        putloc = getloc = 0;
    }
    public void put (char ch)
    {
        if (putloc == q.length - 1)
        {
            System.out.println(" - Queue is full");
            Return;
        }
        putloc ++ ;
        q[putloc] = ch;
    }
    public char get()
    {
        if (getloc == putloc)
        {
            System.out.println(" - Queue is empty.");
            return (char) 0;
        }
        getloc ++ ;
        return q[getloc];
    }
}
```

(4) 给 QueueDemo.java 加上如下的 CircularQueue 类,它实现了字符循环队列:

```java
class CircularQueue implements IQueue
{
    private char q[];
    private int putloc, getloc;
    public CircularQueue(int size)
    {
        q = new char[size + 1];
        putloc = getloc = 0;
    }
    public void put(char ch)
```

```
    {
        if (putloc + 1 = = getloc | ((putloc = = q.length – 1) & (getloc = = 0)))
        {
            System.out.println(" – Queue is full.");
            return;
        }
        putloc ++ ;
        if (putloc = = q.length) putloc = 0;
        q[putloc] = ch;
    }
    public char get()
    {
        if (getloc = = putloc)
        {
            System.out.println(" – Queue is empty");
            return (char) 0;
        }
        getloc ++ ;
        if (getloc = = q.length) getloc = 0;
        return q[getloc];
    }
}
```

循环队列通过重新使用数组的空间工作,循环数组就可以存储无限多个元素。在循环队列中,当保存新元素导致一个未检索的元素被覆盖,而不是当到达内部数组的末尾时,队列是满的。因此,put()必须检验几个条件以确定队列是否是满的。当 putloc 比 getloc 小 1 时,或者 putloc 在数组的结尾而 getloc 在数组的开头时,队列是满的。当 getloc 和 putloc 相等时,队列是空的。

(5) 最后我们创建测试类 QueueDemo 演示两种队列输出结果的不同。

```
public class QueueDemo
{
    public static void main (String args[])
    {
        Queue q1 = new Queue(10);
        CircularQueue q2 = new CircularQueue(10);
        IQueue iQ;
        char ch;
        int i;
        iQ = q1;
        for (i = 0;i<10;i ++ )
```

```java
      iQ.put((char)('A' + i));
    System.out.print("contents of fixed queue");
    for (i = 0; i<10; i++)
    {
      ch = iQ.get();
      System.out.print(ch);
    }
    System.out.println();

     iQ = q2;
    for (i = 0; i<10; i++)
       iQ.put((char)('A' + i));
    System.out.print("contents of circular queue");
    for (i = 0; i<10; i++)
    {
      ch = iQ.get();
      System.out.print(ch);
    }
    System.out.println();

    for (i = 10; i<20; i++)
    {
     iQ.put((char)('A' + i));
    }
    System.out.print("Contents of Circular queue:");
    for (i = 0; i<10; i++)
    {
      ch = iQ.get();
      System.out.print(ch);
    }
    System.out.println("\nStore and consume from circular queue.");
    for (i = 0; i<20; i++)
    {
      iQ.put((char) ('A' + i));
      ch = iQ.get();
      System.out.print(ch);
    }
  }
}
```

最后,程序运行结果如图 4.6 所示。

```
contents of fixed queue:ABCDEFGHIJ
contents of circular queue:ABCDEFGHIJ
Contents of Circular queue:KLMNOPQRST
Store and consume from circular queue.
ABCDEFGHIJKLMNOPQRST
```

图 4.6　测试类 QueueDemo 运行结果

## 4.5　实验习题

1. 选择题

(1) 下列哪些定义抽象类的语句是合法的?(　　)

A. ```
public class JavaDemo{
    abstract void fly();
}
```
B. ```
public abstract JavaDemo{
    abstract void fly();
}
```
C. ```
public abstract class JavaDemo{
    abstract void fly();
}
```
D. ```
public class abstract JavaDemo{
    abstract void fly();
}
```

(2) 对于 Person、Student 和 Teacher 三个类,如有 Student 和 Teacher 是 Person 的子类,如果有一个方法需要同时支持这 3 种类型的参数,则可以将参数的类型设置成(　　)。

A. Person　　　　B. Student　　　　C. Teacher　　　　D. Object

(3) 当编译和运行下列代码时会发生什么?(　　)

```
class Person{
    private  void  eat (String  food){
        System.out.println("person eat" + food);
    }
}
class Superman extends Person{
    public  void  eat (String  food){
        System.out.println("superman eat" + food);
    }
    public static void main(String args[]){
        Superman sm = new Superman();
        sm.eat("bread");
    }
}
```

A. 编译错误,因为 Person 类中的 eat 方法是 private.

B. 运行错误,因为 Person 类中的 eat 方法是 private。
C. 输出 Person：bread
D. 输出 Superman：bread

(4) 下列抽象方法的定义哪些事正确的？（　　）
A. public abstract jump();   B. public abstract jump(){};
C. public abstract void jump();   D. public abstract void jump(){}

(5) 下面的程序中定义了一个 Java 接口,其中包含(　　)处错误。

```
public interface Utility
{
    private int MAX_SIZE = 20;
    int MIN_SIZE = 10;
      void user(){
        System.out.println("using it");
    }
    private int getSize();
    void setSize(int i);
}
```

A. 1　　　　B. 2　　　　C. 3　　　　D. 4

(6) 给定如下 Java 代码,可以填入下划线处的语句是(　　)(选三项)。

```
public interface Utility{}
class FourWheeler implements Utility{}
class Car extends FourWheeler{}
class Bus extends FourWheeler{}
Public class Test{
    public static void main(String[] args){
        _____
    }
}
```

A. Utility car ＝ new Car();   B. FourWheeler bus ＝ new Bus();
C. Utility ut ＝ new Utility();   D. Bus bus ＝ new FourWheeler();
E. FourWheeler fw ＝ new FourWheeler

(7) 以下(　　)修饰符用于声明一个类变量,且该变量的值在运行期间始终保持不变。
A. static　　　B. final　　　C. static final　　　D. final static

(8) 给定下面的 Java 代码,可以填入下划线处的语句是(　　)。

```
public interface Constants{
    int MAX = 50;
    int MIN = 1;
}
public class Test{
```

```
    public static void main(String[] arrgs){
        _____
    }
}
```

A. Contants con = new Contants();

B. Contants.MAX = 100;

C. int i = Contants.MAX—Contants.MIN;

D. Contants.MIN>0;

（9）公有成员变量 MAX 是一个 int 型值，变量的值保持常数值 100。在接口 Constants 中，可以使用以下（　　）声明语句来定义这个变量。

A. public int MAX =100;　　　　B. final int MAX =100;

C. public static int MAX = 100;　　D. public final int MAX = 100;

（10）给定下面的 Java 代码

```
interface foo {
    int k = 0;
}

public class test implements Foo {
    public static void main(String args[]) {
        int i;
        Test test = new test ();
        i = test.k;
        i = Test.k;
        i = Foo.k;
    }
}
```

下面哪一个是发生的结果：（　　）。

A. Compilation succeeds.

B. An error at line 2 causes compilation to fail.

C. An error at line 9 causes compilation to fail.

D. An error at line 10 causes compilation to fail.

E. An error at line 11 causes compilation to fail.

（11）下面哪一个是 MouseMotionListener 接口中的方法？（　　）

A. public void mouseMoved(MouseEvent)

B. public boolean mouseMoved(MouseEvent)

C. public void mouseMoved(MouseMotionEvent)

D. public boolean MouseMoved(MouseMotionEvent)

E. public boolean mouseMoved(MouseMotionEvent)

2. 编程题

（1）设计一个名为 Person 的接口和它的两个名为 Student 和 Employee 子类。Employee 类又有子类：教员类 Faculty 和职员类 Staff。每个人都有姓名、地址、电话号码和电子邮件地址。学生有班级状态（大一、大二、大三或大四）。将这些状态定义为常量。一个雇员有办公室、工资和受聘日期。定义一个名为 MyDate 的类，包含数据域：year（年）、month（月）和 day（日）。教员有办公时间和级别。职员有职务称号。覆盖每个类中的 toString 方法，显示相应的类名和人员。编写一个测试程序，创建 Person、Student、Employee、Faculty 和 Staff，并且调用他们的 toString() 方法。

（2）定义一个接口 Person，声明相关的属性和方法，再用 Teacher 类和 Student 类去继承这个接口，编写测试类创建 10 个 Teacher 和 10 个 Student 类实例，显示输出相关属性。

**参考文献：**

[1] Herbert Schildt 著 石磊译. 新手学 Java 7 编程. 清华大学出版社. 2012.09.

[2] Y. Deniel Liang 著 李娜译. Java 语言程序设计基础篇. 机械工业出版社. 2011.05.

[3] 李伟，张金辉著. Java 入门经典. 机械工业出版社. 2013.04.

# 实验 5　图形用户界面设计

## 5.1　知识点回顾

1. Swing 概要

(1) AWT

Java 在最初发布的时候,提供了一套抽象窗口工具集 AWT(Abstract Window Toolkit,AWT),该工具集提供了一套与本地图形界面进行交互的接口。AWT 主要内容如图 5.1 所示。

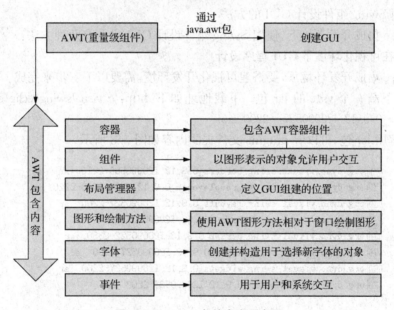

图 5.1　java.awt 包的内容示意图

(2) Swing

Swing 是 AWT 的扩展,是在 AWT 的基础上发展起来的一套新的图形化界面系统,它提供了 AWT 所能提供的所有功能,并且用纯 Java 代码对 AWT 的功能进行了大幅度的扩充。由于 Swing 组件是用纯 Java 代码来实现的,因此在一个平台上设计的组件可以在其他平台上使用且具有相同的效果。由于在 Swing 中没有使用本地方法来实现图形功能,通常把 Swing 组件称为轻量级组件。Swing 组件类一般位于 java 扩展包 javax.swing 中,其中命名一般都以 J 字母开头,如 JFrame、JButton 等。Swing 的主要内容如图 5.2 所示。

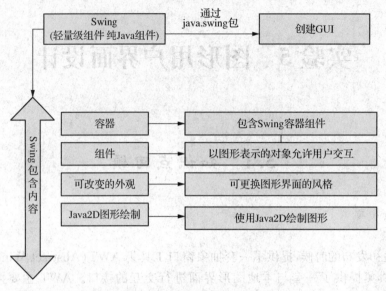

**图 5.2　javax.swing 包的内容示意图**

(3) 使用 Swing 组件设计 GUI 的方法

在 Eclipse 集成开发环境下,使用 Swing 组件设计 GUI 主要有两种方法:

**方法一:在可视化环境下 GUI 程序设计**

在 Eclipse 集成开发环境下,要搭建可视化开发环境,需要以下三步来完成。

第一步,下载一个 vs4e 的 jar 包。下载地址如下:http://visualswing4eclipse.googlecode.com/files/vs4e_0.9.12.I20090527—2200.zip。

下载后解压 jar 包,可以看 plugins 文件夹的内容如图 5.3 所示。

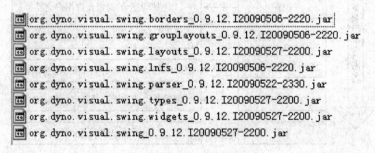

**图 5.3　vs4e 解压包**

将这些内容复制到 Eclipse 安装目录下对应的 plugins 文件夹下,重启 Eclipse。

第二步,启动组件面板。Windows→ShowView→Other→Visual Swing→Palette。如图 5.4 所示。

实验5 图形用户界面设计

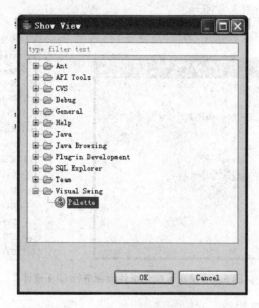

图 5.4 启用可视化组件面板

第三步，使用可视化组件拖拽方式设计 GUI。

新建一个 java 工程 SwingTest，点击工程名右键→新建→other→Frame→点击 next，取包名和类名→finish。如图 5.5 所示。

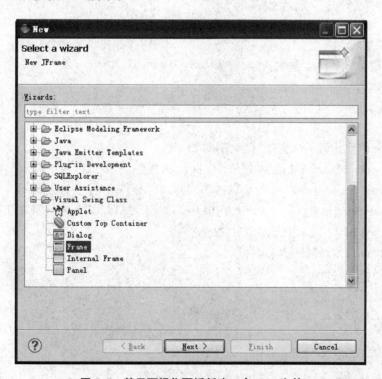

图 5.5 基于可视化面板新建一个 class 文件

我们可以看到出现了两个新的可移动面板，Palette 和 Properties。Swing 各个组件和容器可以通过拖拽的方式来进行布局，组件的外观可以通过对 Properties 面板中的属性进行设置从而改变组件外观。如图 5.6 所示。

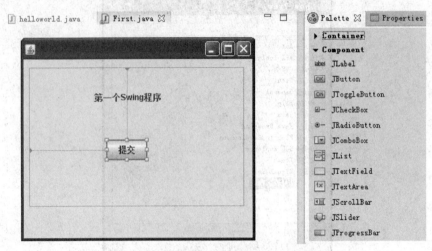

图 5.6　使用可视化组件面板进行 GUI 设计

**方法二：直接编码生成组件**

在源代码文件中，直接利用组件类声明组件，利用组件类的构造方法实例化组件，调用绘图方法在窗口上绘制组件以及具体设置组件的其他属性。比如一个按钮组件的创建过程如下：

```java
import java.awt.*;
import javax.swing.*;
public class buttonFirst extends Frame
{
    private static final long serialVersionUID = 1L;
    JButton btn;
    public buttonFirst()
    {
        super("我的第一个窗口");//窗口 title
        btn = new JButton();//构造一个 JButton 按钮
        btn.setBounds(new Rectangle(226,334,74,30));//绘制按钮
        btn.setFont(new java.awt.Font("Dialog",Font.PLAIN,12));//设置按钮的字号、字体
        btn.setText("提交");//设置按钮文本
        add(btn);//添加至内容面板
        setVisible(true);//设置窗口可见
    }
    public static void main(String[] args)
    {
        buttonFirst btnF = new buttonFirst();
        btnF.setBounds(20, 20, 80, 90);//设置窗体大小
    }
}
```

## 2. 常用 Swing 容器及组件

Swing 组件从功能上可分为以下几种:

顶层容器:JFrame、JApplet、JDialog 和 JWindow 共 4 个。

中间容器:JPanel、JScrollPane、JSplitPane、JToolBar 和 JTabbedPane。

特殊容器:在 GUI 上起特殊作用的中间层,如 JInternalFrame、JLayeredPane 和 JRootPane。

基本控件:实现人机交互的部件,如 JButton、JComboBox、JList、JMenu、JSlider 和 JTextField。

接下来我们主要对一些常用的容器和组件进行功能说明和方法介绍。

(1) 常用 Swing 容器

◆ JFrame

JFrame 是提供给 java 应用程序用来放置图形用户界面的一个容器。Swing 包中的 JFrame 类与我们讲解的 AWT 包中的 Frame 类都与创建窗口有关,JFrame 类是从 Frame 类派生的,包含的方法如表 5.1 所示。

**表 5.1　Jframe 方法说明**

| 方法 | 说明 |
| --- | --- |
| JFrame() | 创建没有标题的窗口 |
| JFrame(GraphicsConfiguration gc) | 创建屏幕设备为 gc 的空标题窗口 |
| JFrame(String title) | 创建以 title 为标题的窗口 |
| Container getContentPane() | 获得窗口的 ContentPane 组件 |
| int getDefaultCloseOperation() | 返回用户关闭窗口时默认的执行操作 |
| void setDefaultCloseOperation() | 设置用户关闭窗口时发生的操作 |
| void update(Graphics g) | 引用 paint()方法重绘窗口 |
| void remove(Component component) | 将窗口中指定的组件删除 |
| JMenuBar getJMenuBar() | 获得窗口中的菜单栏组件 |
| void setJMenuBar(JMenuBar mb) | 设置此窗体上的菜单 |
| void setLayout(LayoutManager manager) | 设置窗口的布局 |

◆ JPanel

面板(JPanel)是一个轻量容器组件,用法与 Panel 相同,用于容纳界面元素,以便在布局管理器的设置下可容纳更多的组件,实现容器的嵌套。Jpanel,JscrollPane,JsplitPane,JinteralFrame 都属于常用的中间容器,是轻量组件。Jpanel 的缺省布局管理器是 FlowLayout。

(2) 常用 Swing 组件

◆ JLabel

该组件提供可带图形的标签,用来显示文字、图标或同时显示文字和图标,包含的方法如表 5.2 所示。

表 5.2　JLabel 方法说明

| 方法 | 说明 |
| --- | --- |
| JLabel() | 创建无图像并且其标题为空字符串的标签实例 |
| JLable(Icon image，int horizontalAlignment) | 创建具有指定图像和水平对齐方式的标签实例 |
| JLabel(Icon image) | 创建具有指定图像的标签实例 |
| JLabel(String text) | 创建具有指定文本的标签实例 |
| JLabel(String text，Icon icon，int horizontalAlignment) | 创建具有指定文本、图像和水平对齐方式的标签实例 |
| JLabel(String text，int horizontalAlignment) | 创建具有指定文本和水平对齐方式的标签实例 |
| getText() | 返回该标签显示的文本字符串 |
| setText(String text) | 定义此组件将要显示的单行文本 |
| setIcon(Icon icon) | 定义此组件将要显示的图标 |
| setFont(Font f) | 设置字体 |
| setForeground(Color c) | 设置当前颜色 |

◆ JTextField

JTextField 类是专门用来建立文本框的,即 JTextField 创建的一个对象就是一个文本框。用户可以在文本框中输入单行的文本,包含的方法如表 5.3 所示。

表 5.3　JTextField 方法说明

| 方法 | 说明 |
| --- | --- |
| JTextField() | 构造一个新的空文本框 |
| JTextField(Document doc，String text，int columns) | 构造一个新的空文本框,它使用给定文本存储模型和给定的列数 |
| JTextField(int columns) | 构造一个具有指定列数的新的空文本框 |
| JTextField(String text) | 构造一个用指定文本初始化的新的文本框 |
| JTextField(String text，int columns) | 构造一个用指定文本和列初始化的新的文本框 |
| setText(String s) | 设置文本框中的文本为参数 s 指定的文本,文本框中原有内容被清除 |
| getText() | 获得文本框中的文本 |
| setEditable(boolean b) | 设定文本框的可编辑性,默认是可编辑的 |
| getColumns() | 返回文本字段中的列数 |
| setColumns(int columns) | 设置文本字段中的列数,然后使布局无效 |
| setHorizontalAlignment(int value) | 设置文本字段中文本的水平对齐方式:JTextField.LEFT |

◆ JTextArea

JTextArea 组件接受用户输入的多行文本,允许用户编辑已输入的文本,包含的方法如表 5.4 所示。

表 5.4 JTextArea 方法说明

| 方法 | 说明 |
| --- | --- |
| JTextArea() | 构造一个新的文本域 |
| JTextArea(Document doc) | 构造一个新的文本域,使其具有给定的文档模型,所有其他参数均默认为(null,0,0) |
| JTextArea(Document doc, String text, int rows, int columns) | 构造具有指定行数和列数以及给定模型的新的文本域 |
| JTextArea(int rows, int columns) | 构造具有指定行数和列数的新的文本域 |
| JTextArea(String text) | 用指定的显示文本构造一个新的文本区 |
| setFont(Font f) | 设置文本区的字体 |
| getText() | 获取文本区中的文本字符串 |

◆ JButton

JButton 类派生自 javax.swing.AbstractButton 类,该类由 JComponent 扩展而来,JButton 对象包含一个文本标签、图像图标或两者,描述按钮、文本/图标周围的空白区域和边框的用途,包含的方法如表 5.5 所示。

表 5.5 JButton 方法说明

| 方法 | 说明 |
| --- | --- |
| JButton() | 创建不带有设置文本或图标的按钮 |
| JButton(Action a) | 创建一个按钮,其属性从所提示的 Action 中获取 |
| JButton(Icon icon) | 创建一个带图标的按钮 |
| JButton(String text) | 创建一个带文本的按钮 |
| JButton(String text, Icon icon) | 创建一个带初始文本和图标的按钮 |
| setLabel(String s) | 设置按钮上的显示文字 |
| getLabel() | 获得按钮上的显示文字 |
| setRolloverIcon(Icon img) | 当鼠标经过时,显示指定的图标 |
| setSelectedIcon(Icon img) | 当选择按钮时,显示 img 指定的图标 |

◆ JCheckBox

JCheckBox 类是专门用来建立复选按钮的,即 JCheckBox 创建的一个对象就是一个复选按钮。复选按钮有两种状态,一种是选中,另一种是未选中,包含的方法如表 5.6 所示。

表 5.6 JCheckBox 方法说明

| 方法 | 说明 |
| --- | --- |
| JCheckBox() | 创建没有文本、没有图标并且最初未被选定的复选框 |
| JCheckBox(Action a) | 创建一个复选框,其属性从所提供的 Action 获取。 |
| JCheckBox(Icon icon) | 创建一个有图标,最初未被选定的复选框 |

(续表)

| 方法 | 说明 |
| --- | --- |
| JCheckBox(Icon icon,boolean selected) | 创建一个带有图标的复选框,并指定其最初是否处于选定状态 |
| JCheckBox(String text) | 创建一个带文本的、最初未被选定的复选框 |
| JCheckBox(String text, boolean selected) | 创建一个带文本的复选框,并指定其最初是否处于选定状态 |
| JCheckBox(String text,Icon icon) | 创建带有指定文本和图标的、最初未被选定的复选框 |
| JCheckBox(String text, Icon icon, boolean selected) | 创建一个带文本和图标的复选框,并指定其最初是否处于选定状态 |
| setSelectedIcon(Icon img) | 创建没有文本、没有图标并且最初未被选定的复选框 |

◆ JRadioButton

JRadioButton 类是专门用来建立单选按钮的,即 JRadioButton 创建的一个对象就是一个单选按钮,包含的方法如表 5.7 所示。

表 5.7 　 JRadioButton 方法说明

| 方法 | 说明 |
| --- | --- |
| JRadioButton() | 创建一个初始化为未选择的单选按钮,其文本未设定 |
| JRadioButton(Action a) | 创建一个单选按钮,其属性来自提供的 Action |
| JRadioButton(Icon icon) | 创建一个初始化为未选择的单选按钮,其具有指定的图像但无文本 |
| JRadioButton(Icon icon, boolean selected) | 创建一个具有指定图像和选择状态的单选按钮 |
| JRadioButton(String text) | 创建一个具有指定文本的状态为未选择的单选按钮 |
| JRadioButton(String text,boolean selected) | 创建一个具有指定文本和选择状态的单选按钮 |
| JRadioButton(String text, Icon icon) | 创建一个具有指定的文本和图像并且初始化为未选择的单选按钮 |
| JRadioButton(String text, Icon icon, boolean selected) | 创建一个具有指定文本、图像和选择状态的单选按钮 |
| getText( ) | 获得单选按钮的标题名 |
| setText(String str) | 设置单选按钮的标题名为 str |

◆ JComboBox

在 Swing 中,组合框由 JComboBox 类表示。ComboBox 是文本字段和下拉列表的组合,让用户可以键入值或从显示给用户的值中进行选择,包含的方法如表 5.8 所示。

表 5.8 　 JcomboBox 方法说明

| 方法 | 说明 |
| --- | --- |
| JcomboBox() | 创建具有默认数据模型的组合框 |
| JcomboBox(ComboBoxModel aModel) | 创建一个组合框,其项取自现有的 ComboBoxModel 中 |
| JcomboBox(Object[] items) | 创建包含指定数组中的元素的组合框 |

(续表)

| 方法 | 说明 |
| --- | --- |
| JcomboBox(Vector<?> items) | 创建包含指定 Vector 中元素的组合框 |
| addItem(Object obj) | 将项添加至项的列表 |
| getItemAt(int index) | 返回指定索引位置的列表项 |
| getItemCount() | 返回列表(作为对象)中的项数 |
| getSelectedItem() | 将当前选择的项作为一个对象返回 |
| getSelectedIndex() | 返回当前选择项的索引位置 |

◆ JPasswordField

JPasswordField 类是专门用来建立密码文本框的,即 JPasswordField 创建的一个对象就是一个密码文本框。用户可以在文本框中输入密码文本,包含的方法如表 5.9 所示。

表 5.9　JPasswordField 方法说明

| 方法 | 说明 |
| --- | --- |
| JPasswordField() | 构造一个新密码框,使其具有默认文档、为 null 开始文本字符串和为 0 的列宽度 |
| JPasswordField(Document doc, String text, int columns) | 构造一个使用给定文本存储模型和给定列数的新密码框 |
| JPasswordField(int columns) | 构造一个具有指定列数的新的空密码框 |
| JPasswordField(String text) | 构造一个利用指定文本初始化的新密码框 |
| JPasswordField(String text, int columns) | 构造一个利用指定文本和列初始化的新密码框 |
| getPassword() | 获得组件中的密码文本 |
| getEchoChar() | 获得隐藏密码所设置的字符 |
| setEchoChar(char c) | 设置隐藏密码而显示的字符为 c,默认为"*" |

◆ JList

JList 类是专门用来建立列表的,即 JList 创建的一个对象就是一个列表。列表中的每项内容可以任意,不局限于 String。它支持单选和多选,包含的方法如表 5.10 所示。

表 5.10　JList 方法说明

| 方法 | 说明 |
| --- | --- |
| JList() | 构造一个使用空模型的列表 |
| JList(ListModel dataModel) | 构造一个列表,使其使用指定的非 null 模型显示元素 |
| JList(Object[] listData) | 构造一个列表,使其显示指定数组中的元素 |
| JList(Vector<?> listData) | 构造一个列表,使其显示指定 Vector 中的元素 |
| setVisibleRowCount(int n) | 设置列表可见行数 |
| setFixedCellHeight(int h) | 设置列表框的固定高度(像素) |
| setFixedCellWidtht(int w) | 设置列表框的固定宽度(像素) |
| isSelectedIndex(int index) | 判断索引为 index 的项是否被选中 |

3. 菜单组件

一般菜单格式包含有菜单栏(JMenuBar)类、菜单(JMenu)类和菜单项(JMenuItem)类对象组成。

(1) 菜单栏

菜单栏是用来管理菜单，不参与交互操作。菜单栏是由 JMenuBar 类派生，菜单栏至少有一个菜单组件才会在图形界面上显现出来。

(2) 菜单

菜单是用来存放菜单项和整合菜单项的组件，菜单是由 JMenu 派生。菜单可以是单一层次菜单，也可以是多层次的结构。

(3) 菜单项

菜单项是菜单系统中最基本的组件，它是由 JMenuItem 类派生。从继承关系来看，菜单项继承了 AbstractButton 类，因此 JMenuItem 具有许多 AbstractButton 类的特性，所以 JMenuItem 支持按钮功能，当选择了菜单项就如同单击某个按钮一样会触发 ActionEvent 事件。这样可以通过 ActionListener 对不同的菜单项编写相应的程序代码。

菜单栏、菜单、菜单项类的方法介绍如表 5.11 所示。

表 5.11 菜单组件方法说明

| JMenuBar 类 | 功能说明 |
| --- | --- |
| JMenuBar() | 创建菜单栏 |
| JMenu() | 创建菜单 |
| JMenu(String str) | 创建具有指定文字的菜单 |
| JMenu(String str,bolean b) | 创建具有指定文字的菜单，通过布尔值确定它是否有下拉式菜单 |
| JMenu 类方法 | 功能说明 |
| JMenuItem(JMenuItem menuitem) | 将菜单项添加到菜单的末尾 |
| Void addSeparator() | 在菜单末尾添加一条分隔线 |
| JMenuItem 类 | 功能说明 |
| JMenuItem() | 创建一个菜单项 |
| JMenuItem(String str) | 创建具有指定文字的菜单项 |
| JMenuItem(Icon icon) | 创建具有指定图形的菜单项 |
| JMenuItem(String str,Icon icon) | 创建具有指定文字和图形的菜单项 |
| JMenuItem(String str,int nmeminic) | 创建一个指定标签和键盘设置快捷键的菜单项 |

制作菜单的一般步骤：

创建一个 JMenuBar 对象并将其加入到 JFrame 中。

创建 JMenu 对象。

创建 JMenuItem 对象并将其添加到 JMenu 对象中。

将 JMenu 对象添加到 JMenuBar 中。

4. 对话框组件

对话框是向用户显示信息并获取程序运行所需数据的窗口，可以起到与用户交互的作用。

Swing 使用 JOptionPane 类提供许多现成的对话框，如：消息对话框、确认对话框、输入对话框等等。如果 JOptionPane 提供的对话框不能满足需要，可以使用 JDialog 类自行设计对话框。JOptionPane 对话框分为 4 种类型：

(1) showMesageDialog：向用户显示一些消息。

(2) showConfirmDialog：问一个要求确认的问题并得到 yes/no/cancel 响应。

(3) showInputDialog：提示用户进行输入。

(4) showOptionDialog：可选择的对话框，该对话框是前面几种形态的综合。

5. 布局

和 AWT 相同，为了容器中的组件能实现平台无关的自动合理排列，Swing 也采用了布局管理器来管理组件的排放、位置、大小等布置任务，在此基础上将显示风格做了改进。

Swing 虽然有顶层容器，但是我们不能把组件直接加到顶层容器中，Swing 窗体中含有一个称为内容面板的容器(ContentPane)，在顶层容器上放内容面板，然后把组件加入到内容面板中。常用的布局管理器主要有流式布局管理器(FlowLayout)、边界布局管理器(BorderLayout)和网格布局管理器(GridLayout)。

(1) 流式布局管理器(FlowLayout)

其特点是在一行上水平排列组件，直到没有足够的空间为止，再开始新的一行。用户缩放容器时，布局管理器自动地调整组件的位置使其填充可用的空间，但容器中组件的大小不会变。面板(JPanel)的默认布局管理器就是流式布局管理；默认情况下，组件是在一行上居中显示。程序员可以设置容器中的组件按左对齐或者右对齐的方式排列。

(2) 边界布局管理器(BorderLayout)

边界布局是将整个容器分为中部、北部、南部、东部或者西部五个区域，程序员可以选择将组件放在任何一个区域，默认放在中部。JFrame 的内容窗格的默认布局管理器就是边界布局。例如：

frame. add(button,BorderLayout. SOUTH)；

边界布局管理器在安放组件时，会先放入边缘组件，剩余的可用空间由中间组件占用，如果有某个边缘组件空缺，其他组件会填充该边缘位置。容器缩放时，边界布局会扩大所有组件的尺寸以便填充可用空间，但边缘组件的厚度不会改变，长度会有所改变，而中间组件的大小会发生变化。

(3) 网格布局管理器(GridLayout)

将容器按行列平均划分；容器上的组件添加从第一行的第一列开始，然后是第一行的第二列，以此类推；

Java.awt. GridLayout

GridLayout( )；

GridLayout(int rows, int cols)；

GridLayout(int rows, int cols, int hgap, int vgap)；

(4) 绝对布局

使用布局管理器时，布局管理器负责各个组件的大小和位置，因此用户无法在这种情况下设置组件大小和位置属性，如果试图使用 Java 语言提供的 setLocation( )、setSize( )、setBounds( )等方法，则都会被布局管理器覆盖，起不到作用。如果开发者确实需要亲自设置组件大小或位置，则应取消该容器的布局管理器，方法为：setLayout(null)。此处所谓的绝对布局，实际上是利用 Java 提供的 setLocation( )、setSize( )、setBounds( )方法，将组件位置固定下来,以达到用户"随心所欲"的去设置布局位置的效果。综合练习即是使用该布局方法。

6. 事件处理

图形化用户界面操作通常是通过鼠标和键盘操作来实现的。鼠标和键盘的操作会引起某一个事件,用户编程识别和处理这些事件。例如,单击图形用户界面上的某个按钮,为接收用户的这一命令,系统应该先识别这一事件,接收该事件后做出相应的响应。在图形化界面的操作中,当某种事件发生时,系统将自动调用预先定义好的相应事件的代码,及时地对发生的事件做出相应处理。事件产生和处理流程如图 5.7 所示。

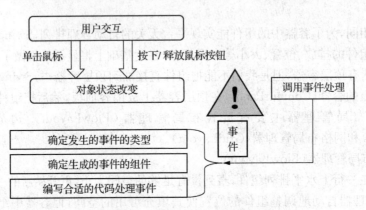

**图 5.7 事件响应处理过程**

(1) 事件类

用户对界面操作在 Java 语言上的描述,以类的形式出现。Java 根据不同的用户操作,产生不同的事件(Event)类。Java.util.EventObject 类是所有事件对象的基础父类,所有事件都是由它派生出来的。不同 EventObject 子类代表不同类型的事件并提供关于该事件的具体信息,这些事件包含在 Java.awt.event 和 Java.swing.event 包中。例如键盘操作对应的事件类是 KeyEvent。常用的事件类如表 5.12 所示。

**表 5.12 常用事件类**

| 事件类 | 相关事件说明 |
| --- | --- |
| ActionEvent | 动作事件:按钮按下,TextField 中按 Enter 键,双击列表项或选择菜单 |
| AdjustmentEvent | 调节事件:在滚动条上移动滑块以调节数值 |
| ComponentEvent | 组建事件:组建尺寸的变化、移动 |
| FocusEvent | 焦点事件:焦点的获得和丢失 |
| ItemEvent | 项目事件:单击复选框和列表项时 |
| WindowEvent | 窗口事件:关闭窗口,窗口闭合,图标化 |
| TextEvent | 文本事件:更改文本框中的信息时 |
| MouseEvent | 鼠标事件:鼠标单击、移动、拖拽 |
| KeyEvent | 键盘事件:键按下、释放 |

(2) 事件源

事件源就是事件发生的场所,通常就是各个组件,例如按钮。

（3）事件监听器

事件监听器实现专门的监听接口对象，接收事件对象，并对其进行处理。常用的监听器如表 5.13 所示。

表 5.13 事件和监听器说明

| 事件类 | 时间监听器（接口） | 监听器描述 |
| --- | --- | --- |
| ActionEvent | ActionListener | 定义了一个接收动作事件的方法 |
| AdjustmentEvent | AdjustmentListener | 定义了一个接收调整事件的方法 |
| ComponentEvent | ComponentListener | 定义了 4 个方法来识别何时隐藏、移动、改变大小、显示组件 |
| FocusEvent | FocusListener | 定义了两个方法来识别何时组件获得或失去焦点 |
| ItemEvent | ItemListener | 定义了一个方法来识别何时项目状态改变 |
| WindowEvent | WindowListener | 定义了七个方法来识别何时窗口激活、关闭、失效、最小化、还原、打开和退出 |
| TextEvent | TextListener | 定义了一个方法来识别何时文本值改变 |
| MouseEvent | MouseListener MouseMotionListener | 定义了两个方法来识别何时鼠标拖动和移动 |
| KeyEvent | KeyListener | 定义了 3 个方法来识别何时键按下、释放和输入字符事件 |

在 Java 的事件机制中，将所有用户对界面的操作定义为事件类，一个事件类对应一类事件。每个事件类都定义了一个或者多个监听器接口，在监听器接口中声明了该事件的各种处理方法。定义一个监听器类来实现相应事件类的监听器接口，给相应组件注册上监听器（即监听器类的实例）。

## 5.2 实验练习

### 5.2.1 实验任务 1

● 实验任务

通过手工编写代码的方式设计一个简单的计算器模块图形界面。运行结果如图 5.8 所示。

图 5.8 简单计算器运行结果

● 实验要点
1. 了解 AWT 和 Swing 的区别与联系。
2. 掌握基本的 Swing 组件和容器使用方法。
3. 学会使用三种常用布局方法灵活布局。
4. 能够运用常用 Swing 组件编写 Java 应用程序的 GUI。

● 实验分析

要实现如图 5.8 所示的计算器的这种布局,需要用到两种布局管理器:BorderLayout 和 GridLayout。一个用于显示计算结果的按钮,使用 BorderLayout 放置于 North;另外 16 个按钮放到一个子面板,置于根面板的 Center,该子面板内使用 GridLayout 布局。由于按钮相对比较多,为了减少相同代码的重复编写,最好使按钮数组通过循环来实现按钮的初始化和添加等工作。

● 实验代码

```java
import java.applet.Applet;
import java.awt.*;
import javax.swing.*;
import javax.swing.JTextField;
public class CalculatorLayout extends Applet
{
    private static final long serialVersionUID = 1L;
    //定义成员变量
    JButton btns[];
    String btnName[] = { "7", "8", "9", "/", "4", "5", "6", "*", "1", "2", "3", "-", "0", "=", "C", "+" };
    JTextField displayText;
    // 定义外框架布局
    BorderLayout Bl;
    // 定义按钮排列布局
    GridLayout Gl;
    JPanel pan1;
    // 初始化各个组件并加入容器
    public void init()
    {
        pan1 = new JPanel();
        displayText = new JTextField(20);
        displayText.setText("0");
        btns = new JButton[btnName.length];
        for (int i = 0; i < btnName.length; i++)
        {
```

```
            btns[i] = new JButton(btnName[i]);
            pan1.add(btns[i]);
        }
        Bl = new BorderLayout();
        setLayout(Bl);
        Gl = new GridLayout(4, 3, 3, 3);
        pan1.setLayout(Gl);
        add(displayText, "North");
        add(pan1, "Center");
        add(pan1);
    }
}
```

### 5.2.2 实验任务 2

● 实验任务

实验 5.2.1 只是完成的简单计算器的外观布局,本实验任务是实现加、减、乘、除和清零等功能,完成一个简单计算器模块。

● 实验要点

1. 理解 Java 事件处理机制。
2. 掌握 Java 中常用的事件类和常用事件监听器。
3. 能够编写常用组件的事件处理程序。

● 实验分析

实验 5.2.1 只是完成的简单计算器的界面设计,要实现加减乘除和清零的功能,首先要考虑两个方面:第一,大于 9 的数字,比如 111,怎么输入?那就需要判定将要点击的下一个按钮是数字键还是运算键,这就需要一个变量 change 用来存储当前点击的这个键是否是运算键。第二,在进行运算时,当点击了运算符,如"+"号,那么前一个数据需要保存在一个变量里,否则的话,再点击运算符后的另外一个数据时,前一个数据已经被覆盖丢失了,从而也不能进行运算了。所以,要完成计算器的功能,需要几个用来存储数据和状态的变量。

运行结果如图 5.9 所示。

图 5.9　简单计算器功能实现运行结果

● **实验代码**

```java
import java.applet.Applet;
import java.awt.*;
import java.awt.event.ActionEvent;
import java.awt.event.ActionListener;
import javax.swing.*;
import javax.swing.JTextField;
public class Calculator extends Applet implements ActionListener
{
    private static final long serialVersionUID = 1L;
    // 定义成员变量(可看做全局变量)
    // 暂存显示数据(当前显示的数据或者点击运算符号以后的数据)
    String tempText = "0";
    // 暂存符号
    String sign = null;
    // 暂存内部运算数据(点击运算符号之前的数据)
    double tempData = 0;
    boolean change = false;
    // 暂存是否已经有了内部运算(即是否点了运算符号按钮)
    JButton btns[];
    String btnName[] = { "7", "8", "9", "/", "4", "5", "6", "*", "1", "2", "3", "-", "0", "=", "C", "+" };
    JTextField displayText;
    // 定义外框架布局
    BorderLayout Bl;
    // 定义按钮排列布局
    GridLayout Gl;
    JPanel pan1;
    public void init()
    {
        //初始化各个组件并添加监听器
        pan1 = new JPanel();
        displayText = new JTextField(20);
        displayText.setText("0");
        btns = new JButton[btnName.length];
        for (int i = 0; i < btnName.length; i++)
        {
            btns[i] = new JButton(btnName[i]);
```

```java
            btns[i].addActionListener(this);
            pan1.add(btns[i]);
        }
        Bl = new BorderLayout();
        setLayout(Bl);
        Gl = new GridLayout(4, 3, 3, 3);
        pan1.setLayout(Gl);
        add(displayText, "North");
        add(pan1, "Center");
        add(pan1);
    }
    public void actionPerformed(ActionEvent e)
    {
        if (e.getSource() == btns[0]) {
press_7();
        } else if (e.getSource() == btns[1]) {
            press_8();
        } else if (e.getSource() == btns[2]) {
            press_9();
        } else if (e.getSource() == btns[3]) {
            press_div();
        } else if (e.getSource() == btns[4]) {
            press_4();
        } else if (e.getSource() == btns[5]) {
            press_5();
        } else if (e.getSource() == btns[6]) {
            press_6();
        } else if (e.getSource() == btns[7]) {
            press_mul();
        } else if (e.getSource() == btns[8]) {
            press_1();
        }
        else if (e.getSource() == btns[9]) {
            press_2();
        } else if (e.getSource() == btns[10]) {
            press_3();
        } else if (e.getSource() == btns[11]) {
            press_sub();
        } else if (e.getSource() == btns[12]) {
```

```java
            press_0();
        } else if (e.getSource() == btns[13]) {
            press_equal();
        } else if (e.getSource() == btns[14]) {
            press_clear();
        } else if (e.getSource() == btns[15]) {
            press_add();
        }
    }
    public void press_0()
    {
        if (tempText != "0"&& change == false) {
            tempText = tempText + "0";
        } else if (tempText == "0" || sign != null) {
            tempText = "0";
        }
        displayText.setText(tempText);
    }
    public void press_1()
    {
        if (tempText != "0"&& change == false) {
            tempText = tempText + "1";
        } else if (tempText == "0" || sign != null) {
            tempText = "1";
            change = false;
        }
        displayText.setText(tempText);
    }
    public void press_2()
    {
        if (tempText != "0"&& change == false) {
            tempText = tempText + "2";
        } else if (tempText == "0" || sign != null) {
            tempText = "2";
            change = false;
        }
        displayText.setText(tempText);
    }
    public void press_3()
```

```java
{
    if (tempText ! = "0"&& change = = false) {
        tempText = tempText + "3";
    } else if (tempText = = "0" || sign ! = null) {
        tempText = "3";
        change = false;
    }
    displayText.setText(tempText);
}
public void press_4()
{
    if (tempText ! = "0"&& change = = false) {
        tempText = tempText + "4";
    } else if (tempText = = "0" || sign ! = null) {
        tempText = "4";
        change = false;
    }
    displayText.setText(tempText);
}
public void press_5()
{
    if (tempText ! = "0"&& change = = false) {
        tempText = tempText + "5";
    } else if (tempText = = "0" || sign ! = null) {
        tempText = "5";
        change = false;
    }
    displayText.setText(tempText);
}
public void press_6()
{
    if (tempText ! = "0"&& change = = false) {
        tempText = tempText + "6";
    } else if (tempText = = "0" || sign ! = null) {
        tempText = "6";
        change = false;
    }
    displayText.setText(tempText);
}
```

```java
public void press_7()
{
    if (tempText ! = "0"&& change = = false) {
        tempText = tempText + "7";
    } else if (tempText = = "0" || sign ! = null) {
        tempText = "7";
        change = false;
    }
    displayText.setText(tempText);
}
public void press_8()
{
    if (tempText ! = "0"&& change = = false) {
        tempText = tempText + "8";
    } else if (tempText = = "0" || sign ! = null) {
        tempText = "8";
        change = false;
    }
    displayText.setText(tempText);
}
public void press_9()
{
    if (tempText ! = "0"&& change = = false) {
        tempText = tempText + "9";
    } else if (tempText = = "0" || sign ! = null) {
        tempText = "9";
        change = false;
    }
    displayText.setText(tempText);
}
public void press_add()
{
    sign = "add";
    tempData = Double.parseDouble(tempText);
    change = true;
}
public void press_sub()
{
    sign = "sub";
```

```
        tempData = Double.parseDouble(tempText);
        change = true;
}
public void press_mul()
{
        sign = "mul";
        tempData = Double.parseDouble(tempText);
        change = true;
}
public void press_div()
{
        sign = "div";
        tempData = Double.parseDouble(tempText);
        change = true;
}
public void press_clear()
{
        displayText.setText("0");
        tempData = 0;
        tempText = "0";
}
public void press_equal()
{
    if (sign = = null)
        displayText.setText(tempText);
    if (sign = = "add") {
        tempData + = Double.parseDouble(tempText);
        tempText = Double.toString(tempData);
        displayText.setText(tempText);
    } else if (sign = = "sub") {
        tempData - = Double.parseDouble(tempText);
        tempText = Double.toString(tempData);
        displayText.setText(tempText);
    } else if (sign = = "mul") {
        tempData * = Double.parseDouble(tempText);
        tempText = Double.toString(tempData);
        displayText.setText(tempText);
    } else if (sign = = "div") {
        if (tempData ! = 0) {
```

```
                tempData /= Double.parseDouble(tempText);
                tempText = Double.toString(tempData);
                displayText.setText(tempText);
            } else
                tempText = "0";
            displayText.setText(tempText);
        }
        sign = null;
    }
}
```

## 5.3 项目实战

### 5.3.1 项目描述

本实验项目是使用可视化 GUI 程序设计方式,设计和实现用户注册功能。当点击【提交】按钮时,用户信息显示在窗体上,当点击【重置】按钮时,用户注册信息空间清空。

### 5.3.2 项目分析

在可视化程序设计方式下,要完成以上功能主要分以下三步:
第一,新建窗体,拖拽添加组件、合理布局。
第二,修改组件变量,修改组件属性。
第三,编写两个按钮事件处理程序。

### 5.3.3 项目编写

新建一个 Java 工程,工程名为 SwingTest;新建包名:userLogin。
➢ 新建一个窗体
File→New→other→Visual Swing Class→Frame→输入窗体名为 Login→Finish。我们可以看到一个新的空白窗体被建立,如图 5.10 所示。
窗体右边有可视化组件面板 Palette 和属性面板 Properties。窗体下面是代码区,通过点击窗体下面的两个黑色三角,可以对设计模式和代码模式进行切换。组件面板的组件可以通过点击拖动,拖动到窗体上面,进行设计和组合。空白窗体生成的代码如下:

实验5 图形用户界面设计

图 5.10 建立空白窗体

```
/*导入Swing包*/
import java.awt.*;
import javax.swing.*;
import org.dyno.visual.swing.layouts.Constraints;
import org.dyno.visual.swing.layouts.GroupLayout;
import org.dyno.visual.swing.layouts.Leading;
public class Test extends JFrame
{
    private static final long serialVersionUID = 1L;
    private JLabel jLabel0;
    private JLabel labDisplay;
    private JButton btn;
    private static final String PREFERRED_LOOK_AND_FEEL =
"javax.swing.plaf.metal.MetalLookAndFeel";
    public Test()
```

```
    {
        initComponents();
    }
}
```

➢ 拖拽组件到窗体，并合理布局

从组件面板将需要的主键一个一个拖动到窗体上，并进行合理排版。如图 5.11 所示。

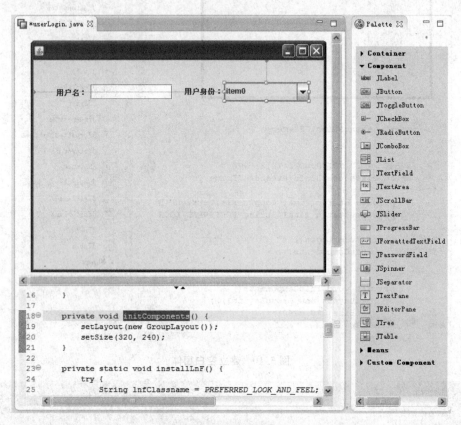

图 5.11　图形化界面设计

➢ 修改组件属性，进行个性化设计

为了在提高程序的可读性和提高编码速度，我们需要对每个组件设置一个有意义的名字。有两种方法可以设置，一个是点击需要修改的组件，找到相应的属性面板，对它的 Bean File Name 的 value 值进行修改；或者点击需要修改名字的组件，右键单击，选择 change variable name 选项，打开修改变量选项框，对组件名称进行修改。如图 5.12 所示。

为了界面更加美观，符合个性化设计，我们可以对每个组件的属性进行修改，比如修改字体颜色、修改组件边框、修改背景色等。选中所需要修改的组件，然后在 Property 面板中进行相应的修改。界面设计完成的效果如图 5.13 所示。

实验 5　图形用户界面设计

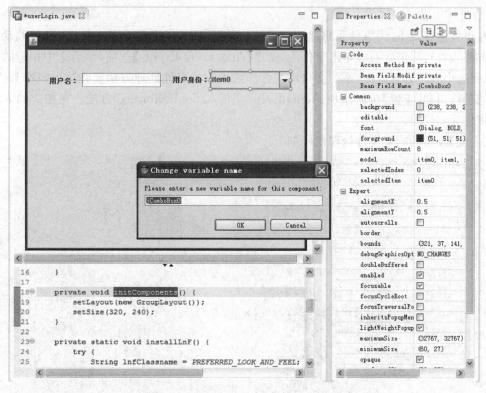

图 5.12　修改组件名称

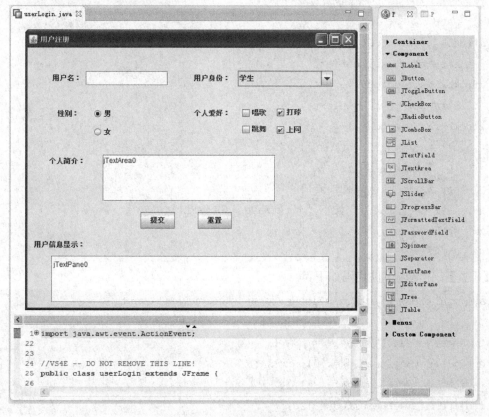

图 5.13　用户注册界面设计完成

➢ 对事件响应进行处理

在可视化方式下,为事件源添加响应事件处理程序入口(即监听接口中定义的接收动作事件的方法入口),只需要右键单击事件源,选中【Add/Edite Events】选项,然后选中相应的事件类,从而选中事件处理方法,系统会自动为你添加相应的事件处理方法。在这里,我们只需要为【提交】按钮和【重置】按钮添加 ActionEvent 类的 actionPerformed 方法。在相应的方法中编写事件处理代码。

【提交】按钮的响应事件处理代码如下:

```
private void btnSubmitActionActionPerformed(ActionEvent event)
{
    str = str + txtName.getText();
    str = str + "";
    if(radioMale.isSelected())
        str = str + radioMale.getText();
    else if(radioFemale.isSelected())
        str = str + radioFemale.getText();
    str = str + " 是一名";
    if(cmbRole.getSelectedIndex()>=0)
    {
        str = str + cmbRole.getSelectedItem();
    }
    str = str + "   爱好   ";
    if(ckb1.isSelected())
        str = str + ckb1.getText() + "";
    if(ckb2.isSelected())
        str = str + ckb2.getText() + "";
    if(ckb3.isSelected())
        str = str + ckb3.getText() + "";
    if(ckb4.isSelected())
        str = str + ckb4.getText() + "";
    str = str + "  个人简介如下:" + txtNates.getText();
    txt.setText(str);
}
```

【重置】按钮响应事件代码如下:

```
private void btnResetActionActionPerformed(ActionEvent event)
{
    txtName.setText("");
    cmbRole.setSelectedIndex(0);
    radioMale.setSelected(false);
    radioFemale.setSelected(false);
```

```
    ckb1.setSelected(false);
    ckb2.setSelected(false);
    ckb3.setSelected(false);
    ckb4.setSelected(false);
    txtNates.setText("");
    txt.setText("");
}
```

运行结果图 5.14 所示。

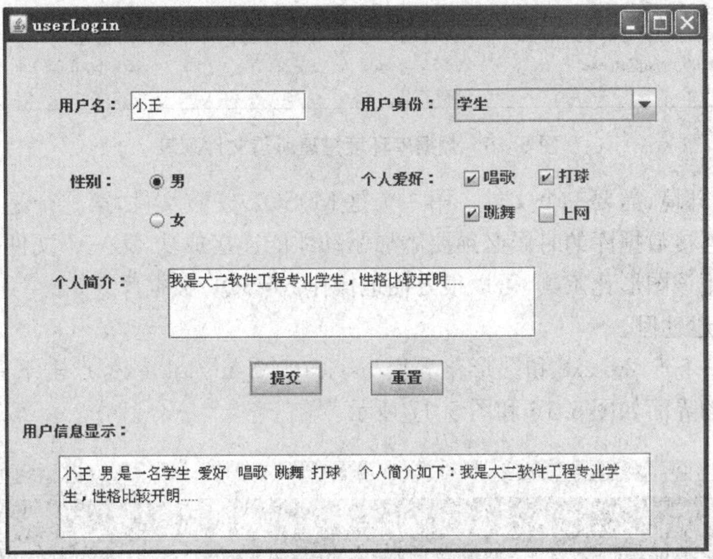

图 5.14 用户注册程序运行结果

## 5.4 综合项目实战

### 5.4.1 项目描述

本实验项目是使用编码方式设计添加组件和布局,完成简易的用户管理小系统。该小系统的主要功能有:用户添加、删除、修改和查询功能。用户删除时,可以根据用户 id 和用户名两种方式来删除,可以根据用户 id 来修改用户,也可以显示所有用户信息。用户信息存储在数据库中。

### 5.4.2 项目分析

(一)要完成该系统,必须要具备以下基础知识:
(1)数据库操作。用 Java 代码实现用户信息的增删改查等功能。

（2）组件的添加和布局。该系统是使用编码方式实现组件的添加，旨在使学生对 Java 类和对象的创建和使用有一个更深刻的理解。

（3）事件处理程序。当用户点击按钮或者菜单时，实现事件处理功能，完成数据的更新操作。

（二）实验步骤：

**第一步，选择数据库，建立数据库和数据表。** 该系统我们选择 MySQL 数据库。

（1）数据库搭建所需要的文件。

| 名称 | 修改日期 | 类型 | 大小 |
|---|---|---|---|
| mysql-5.5.20-win32.zip | 2016-05-16 8:47 | KuaiZip.zip | 30,819 KB |
| mysql-connector-java-5.0.5-bin.jar | 2016-05-16 8:47 | Executable Jar File | 501 KB |
| navicat100_mysql_en.exe | 2016-05-16 8:48 | 应用程序 | 16,545 KB |

图 5.15  数据库环境搭建所需文件列表

搭建数据库环境，需要三个文件。第一个是 MySQL 安装文件，第二个是 MySQL 对应版本的驱动程序，连接数据库的时候必须要添加驱动才能连接成功，第三个文件是 MySQL 图形化界面软件，使用该图形化工具，可以很方便地操作 MySQL 数据库。

（2）环境搭建过程。

安装 mysql-5.5.20.exe 和图形化界面 navicat100_mysql.exe，点击下一步依次完成即可，完成安装后的界面如图 5.16 和图 5.17 所示。

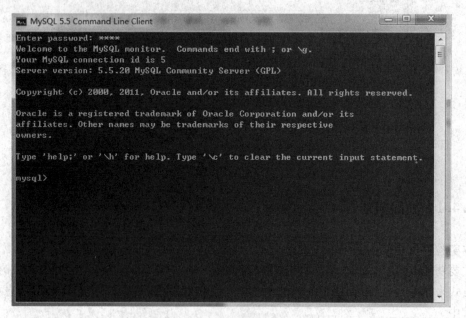

图 5.16  MySQL 数据库启动程序界面

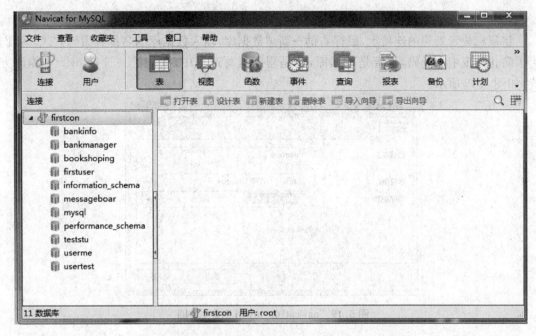

图 5.17 MySQL 图形化界面 navicat100_mysql 启动界面

3) 建立链接

首先,打开 navicat100_mysql,建立连接。点击连接按钮,弹出如下对话框,输入连接名称和用户密码等信息,点击确定完成。如果已经存在一个连接,可以不去建立新连接。该连接实际上是实现与 MySQL 数据库的对接,接下来可以以图形化界面的方式对 MySQL 数据库进行操作。

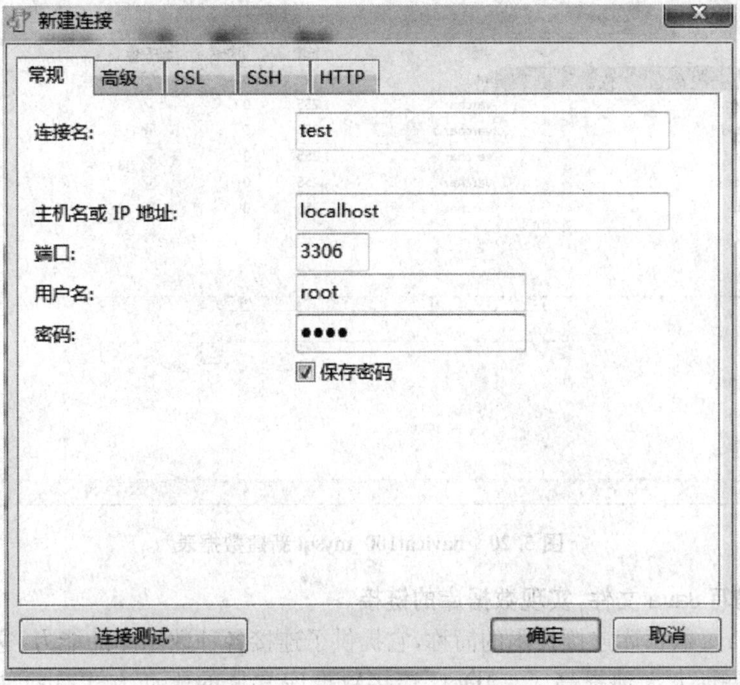

图 5.18 navicat100_mysql 建立链接

4）新建数据库。

将鼠标放置新建的连接上，鼠标右键→新建数据库，输入数据库的名称和编码方式。我们为了防止出现中文乱码，不管是数据库，还是程序编写过程中，我们统一选择 utf-8 编码方式，如图 5.19 所示。

图 5.19  navicat100_mysql 新建数据库

5）新建数据表。

双击新建立的数据库，出现表、函数等列表项，右键表→新建表，建立用户信息表，如图 5.20 所示。其中 id 是用户表的主键，设置成自动增加。至目前为止，数据库的环境已经搭建好，数据库文件也已经建好了。

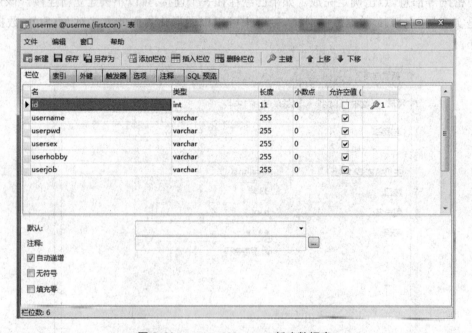

图 5.20  navicat100_mysql 新建数据表

**第二步，编写 Java 文件，实现数据库的链接。**

JDBC 是 Java 数据库连接技术的简称，它提供了连接各种数据库的能力，这便使程序的可维护性和可扩展性大大地提高了。JDBC 连接数据库常见的驱动方式有两种，一种是 jdbc-odbc 即桥连，另外一种是纯 Java 驱动。一般在做 Java 开发的时候用第二种，目前我们使用纯

Java 驱动方式连接数据库。

首先,在 eclipse 里新建一个 Java 工程项目 SwingTest,并在 src 文件夹下,新建一个包,包名为:DB,在该包下新建 Java 文件 DBoper 类。

其次:DBoper 类 连接数据库的方法 getConnection()源码如下:

```
public Connection getConnection() {
    try {
        Class.forName("com.mysql.jdbc.Driver");
        conn = DriverManager.getConnection("jdbc:mysql://localhost:3306/userme?useUnicode=true&characterEncoding=utf8","root", "root");
    } catch (Exception e) {
        System.out.println("数据库连接失败");
    }
    return conn;
}
```

最后,加载 MySQL 驱动程序 mysql-connector-java-5.0.5-bin.jar。

右键工程项目 SwingTest→Built path→Configure Built path→Libraries→Add External Jars,选择驱动名称,加载后如图 5.21 所示。

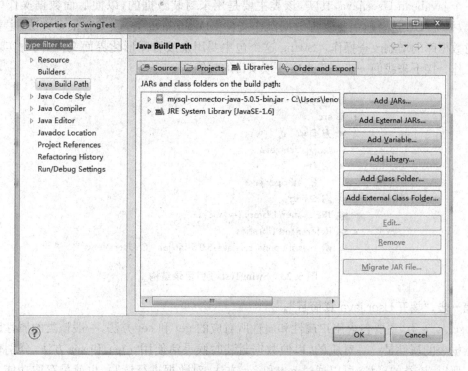

图 5.21　SwingTest 项目加载数据库驱动

加载完驱动以后,会在工程项目文件目录结构中,显示该驱动名称。如图 5.22 所示。

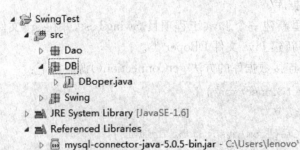

图 5.22  SwingTest 项目驱动加载后所示效果

**第三步**:编写数据库操作文件 DBoper.java 文件。
**第四步**,编写图形化文件,编写事件处理程序。
第三步和第四步具体内容将在 5.4.3 项目编写部分中详细讲解。

### 5.4.3  项目编写

该小系统的主要功能有:用户添加、删除、修改和查询功能。我们先新建一个 Java 工程项目 SwingTest,该工程项目的目录结构如图 5.23 所示。该项目工程有三个包。Dao 这个包下放了一个 javabean User.java,其实,该类主要是用来封装数据的,以便后面数据操作中用来"运载数据"。DB 这个包主要放置数据操作文件 DBoper.java 类,该类有很多方法,主要用来实现对数据库的增删改查操作。Swing 这个包主要用来放置图形化界面文件,同时实现事件处理工作。接下来我们一步一步来实现该项目的功能。

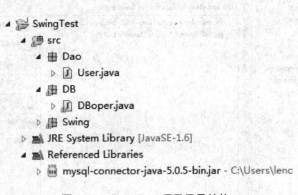

图 5.23  SwingTest 项目目录结构

**第一步**,"揭开 User.java 真面目"。

User.java 类,主要包含用户属性和属性所对应的 get 和 set 方法,一般该类的属性跟数据库里数据表的属性是一一对应的,以便数据封装时候灵活使用。get 和 set 方法,分别代表读和写。所以,该类的对象,可以通过 get 和 set 方法,对数据进行读写,也就是存取功能。这也是 Java 封装性的一种主要体现。下面我们来看一下 User.java 源码。

```
package Dao;
public class User {
    private int userid;
```

```java
public int getUserid() {
    return userid;
}
public void setUserid(int userid) {
    this.userid = userid;
}
private String username;
private String userpwd;
private String usersex;
private String userhobby;
private String userjob;
public String getUserjob() {
    return userjob;
}
public void setUserjob(String userjob) {
    this.userjob = userjob;
}
public String getUsername() {
    return username;
}
public void setUsername(String username) {
    this.username = username;
}
public String getUserpwd() {
    return userpwd;
}
public void setUserpwd(String userpwd) {
    this.userpwd = userpwd;
}
public String getUsersex() {
    return usersex;
}
public void setUsersex(String usersex) {
    this.usersex = usersex;
}
public String getUserhobby() {
    return userhobby;
}
public void setUserhobby(String userhobby) {
```

```
            this.userhobby = userhobby;
     }
}.
```

我们在 main 方法中对 User 类进行简单测试,实现数据的存取功能,即读写功能。

```
public static void main(String[] args) {
    User usertest = new User();//实例化对象,产生一个 usertest 对象
    usertest.setUserhobby("篮球");//set 方法给对象赋值
    usertest.setUserjob("医生");
    usertest.setUsername("张三");
    usertest.setUserpwd("123");
    usertest.setUsersex("男");
    System.out.println(usertest.getUsername());
                           //get 方法把封装的数据取出来,并打印显示出来.
    System.out.println(usertest.getUserpwd());
    System.out.println(usertest.getUsersex());
    System.out.println(usertest.getUserhobby());
    System.out.println(usertest.getUserjob());
}
```

运行结果如图 5.24 所示。

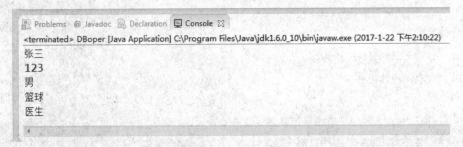

图 5.24　数据封装实例演示

以上是对固定数据的封装。如果对用户注册信息等可变数据的封装,也是可以的。后面会陆续展示。

第二步:"对数据库操作类 DBoper.java 一探究竟"。

DBoper.java 类文件主要包含数据库的链接方法、数据的增加、删除、修改和查询等方法。要深刻理解对数据库的操作,必须了解 JDBC API 中主要的接口和类,如表 5.14 所示。

表 5.14　JDBC API 中主要的接口和类

| 名称 | 解释 |
| --- | --- |
| DriverManager | 处理驱动的调入并且对产生新的数据库连接提供支持 |
| Connection | 代表对特定数据库的连接 |

(续表)

| 名称 | 解释 |
| --- | --- |
| Statement | 代表一个特定的容器,容纳并执行一条 SQL 语句 |
| PreparedStatement | 带有预编译功能的 Statement。预编译语句在创建 PreparedStatement 对象时能将 SQL 语句送给数据库系统进行预编译。当以后要执行同一 SQL 语句时,就直接执行,而不需要先将其编译,从而提高执行速度。 |
| ResultSet | 控制执行查询语句得到的结果集 |

Statement 接口提供了 3 种执行语句的方法,如表所示 5.15 所示。PreparedStatement 也有类似方法。

表 5.15 Statement 主要方法列表

| 方法名称 | 解释 |
| --- | --- |
| executeQuery(String sql) | 用于产生单个结果集的语句,例如,执行 SELECT 查询语句 |
| executeUpdate(String sql) | 用于执行 INSERT、UPDATE 或 DELETE 语句以及 SQL DDL(数据定义语言)语句,例如,CREATE TABLE 和 DROP TABLE。INSERT、UPDATE 或 DELETE 语句的效果是修改表中零行或多行中的一列或多列。executeUpdate 的返回值是一个整数,指示受影响的行数(即更新计数)。对于 CREATE TABLE 或 DROP TABLE 等不操作性的语句,executeUpdate 的返回值总为零。 |
| execute(String sql) | 用于执行返回多个结果集、多个更新计数或二者组合的语句 |

下面我们来看一下,要基于 JDBC 实现对数据库的操作,需要一个怎么样的步骤?

一个基本的 JDBC 程序开发包含以下步骤:
(1) 设置环境,引入相应的 JDBC 类
(2) 选择合适的 JDBC 驱动程序并加载
(3) 分配一个 Connection 对象
(4) 分配一个 Statement 对象
(5) 用该 Statement 对象进行查询等操作
(6) 从返回的 ResultSet 对象中获取相应的数据
(7) 关闭 Connection

首先,依据这个步骤,我们先连接数据库,并测试成功!源码如下:

```
package DB;//新建包名
import java.sql.*;//开发步骤第一步导入 JDBC 类.Java.sql.* 这个包下几乎包含了所以
//JDBC 所需操作类和接口
//第二步骤加载驱动,已经在环境搭建部分讲解,不再赘述
import java.util.ArrayList;
import Dao.User;//导入 User 类
public class DBoper {
    // ******************** 定义常用的四个对象,数据库连接对象 conn、数据库 sql 语句执行
```

```java
//对象 st 和 pstmt,结果集存储对象 rs. ******************
private Connection conn = null;
private Statement st = null;
private PreparedStatement pstmt = null;
private ResultSet rs = null;
private ArrayList<User> result;
// ********************* 连接数据库【编码 UTF-8】**********************
public Connection getConnection() {
    try {
        Class.forName("com.mysql.jdbc.Driver");//驱动名称
        conn = DriverManager.getConnection(
"jdbc:mysql://localhost:3306/userme?useUnicode=true&characterEncoding=utf8","root","root");//连接数据库的字符串.
        System.out.println("数据库连接成功!");

    } catch (Exception e) {
        System.out.println("数据库连接失败");
    }
    return conn;
}
public static void main(String[] args) {
    DBoper op = new DBoper();
    op.getConnection();//调用数据库连接方法,测试是否连接成功
}
}
```

执行 main 方法,如果结果如图 5.25 所示,则说明数据库连接成功。否则,数据库连接失败,就要去排查数据库驱动有没有加载、数据库名称有没有写错等因素。

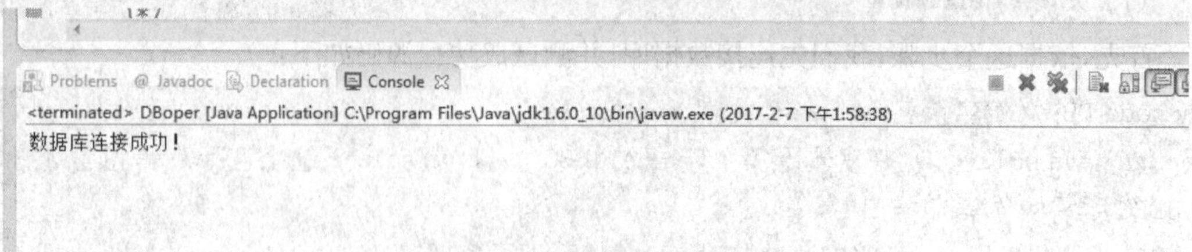

**图 5.25 数据库链接成功**

其次,数据库连接成功后,我们可以对数据库进行操作了。但每个数据库操作完毕后,尽量都要关闭数据库连接,因为数据库连接数量是有限制的,如果随着使用者的增加,连接数量会超出了限制额度,系统就会报错。所以,我们可以写一个关闭数据库的方法,在数据库操作完毕以后,直接调用这个方法来关闭数据库。此方法有三个参数,分别代是 ResultSet 结果集对象 rs,PreparedStatement 对象 pstmt 和 Connection 的 conn 对象。在关闭数据库连接时,一

般情况下直接关闭 Connection 是没有问题，但必须保证的是这里的关闭是真正的关闭，但在使用连接池运用中，关闭 Connection 并不意味着真正的关闭了 Connection，而是将 connection 返回到池中，并没有关闭。在这种情况下，你的 PreparedStatment 和 Resulset 都没有被关闭，所以编码过程中，要养成一个良好的习惯，打开数据时，顺序是：Connection → PreparedStatement→ResultSet，关闭时：ResultSet→PreparedStatement→Connection，关闭数据库连接的方法源码如下所示：

```java
// ****************************** 关闭数据库资源 ******************************
public void releaseDB(ResultSet rs, PreparedStatement pstmt, Connection conn)
{
    if(rs! = null)
    {
        try {
            rs.close();
        } catch (SQLException e) {
            e.printStackTrace();
        }
    }
    if(pstmt! = null)
    {
        try {
            pstmt.close();
        } catch (SQLException e) {
            e.printStackTrace();
        }
    }
    if(conn! = null)
    {
        try {
            conn.close();
        } catch (SQLException e) {
            e.printStackTrace();
        }
    }
}
```

有了数据库的连接和关闭操作，最后，我们来分析数据库的增删改查操作。其实，增加、删除和修改，主要用到 Statement 对象和 PrepareStatement 对象以及他们的 executeUpdate(String sql)方法。查询操作，主要用到 Statement 对象和 PrepareStatement 对象和 ResultSet 对象。Statement 对象和 PrepareStatement 对象的方法 executeQuery(String sql)用来执行查询语句，查询结果存储在 ResultSet 对象中。下面我们主要以 PrepareStatement 对象为主，具体来分析 DBoper 类中数据库操作的各个方法。

1)用 PrepareStatement 对象实现数据的修改和删除数据。具体含义看注释。

```java
// ***************** prepareStatement 修改 删除 数据 **************************
public boolean dataupdate(String sql) {
    conn = getConnection();//连接数据库
    try {
        pstmt = conn.prepareStatement(sql);
                                    //产生 prepareStatement 对象,并传入 sql 参数
        pstmt.executeUpdate();//执行 sql 语句
        System.out.println("操作成功");
    }
    catch (SQLException e) {
        e.printStackTrace();
        System.out.println("操作失败");
    }
    finally
    {
        releaseDB(rs, pstmt, conn);;//调用关闭数据库的方法
    }
    return true;
}
```

2) 用 PrepareStatement 对象实现对某条记录的修改。具体含义看注释。

```java
// ******************************* 依据 id 来修改记录 *******************
public boolean dataupdateid(User user, int userid) {
    conn = getConnection();//连接数据库
    try {
        String sql = "update userme set username = ?, userpwd = ?, usersex = ?, userhobby = ?, userjob = ? where id = " + userid;
                        //sql 语句,?代表参数占位符,具体值在下面调用 set 方法赋值
        pstmt = conn.prepareStatement(sql);
                                    // 产生 prepareStatement,并传入 sql 参数
        pstmt.setString(1, user.getUsername());  //给 sql 语句中的?——赋值,以下同
        pstmt.setString(2, user.getUserpwd());
        pstmt.setString(3, user.getUsersex());
        pstmt.setString(4, user.getUserhobby());
        pstmt.setString(5, user.getUserjob());
        System.out.println(user.getUserjob());
        pstmt.executeUpdate();//执行 sql 语句,实现修改操作
```

```java
            System.out.println("操作成功");
        }
        catch (SQLException e) {
            e.printStackTrace();
            System.out.println("操作失败");
        }
        finally
        {
            releaseDB(rs, pstmt, conn); //关闭数据库
        }
        return true;
}
```

3) 用 PrepareStatement 插入一条记录。具体含义看注释。

```java
// ********************* prepareStatement 插入数据 *************************
public boolean pstmtinsert(User user) { //user 对象是封装了动态数据的一个用户数据对象
    conn = getConnection();
    try {
        pstmt = conn.prepareStatement("insert into userme (username,userpwd,usersex,userhobby,userjob) values(?,?,?,?,?)");
                              //产生 prepareStatement,并传入带参数的 sql 语句
        pstmt.setString(1, user.getUsername()); //给 sql 语句中的?赋值
        pstmt.setString(2, user.getUserpwd());
        pstmt.setString(3, user.getUsersex());
        pstmt.setString(4, user.getUserhobby());
        pstmt.setString(5, user.getUserjob());
        pstmt.executeUpdate(); //执行 sql 语句,插入成功返回 true
    }
    catch (SQLException e) {
        e.printStackTrace();
    }
    finally
    {
        releaseDB(rs, pstmt, conn); //关闭数据库
    }
    return true;
}
```

4) 用 PrepareStatement 查询,返回符合条件的所有记录,可以是一条记录,也可以是多条

记录,并存入 ArrayList 列表里。下面我们来演示存储多条记录除了用 ResultSet 以外,还可以用其他数据结构来保存,比如 ArrayList 列表。思路:每一条用户信息可以存储到一个 user 对象里,多个 user 对象可以存储到 ArrayList 列表里,这样,我们就可以实现多条记录的存储了。需要特别指出的是,有些读者会有疑问,为何不用 ResultSet 的一个对象 rs 来存呢?方法的返回值可以是 rs,反而更简洁。我们主要从两个方面考虑,第一,基于 MVC 的三层架构开发,最基本的宗旨是表现层跟逻辑控制层要分开,如果我们返回结果是 rs,势必在表现层(此处可以认为是图形化界面层)需要导入数据库相关包,并连接数据库,读取数据。而我们如果用当前的 ArrayList 存储,是将有用的数据暂存到一个数据结构中,在表现层可以随时获取。第二,考虑到我们已经编写好的关闭数据库资源的方法 releaseDB(rs, pstmt, conn),如果我们存储到 rs 中,并返回 rs,最后在 finally 里调用 releaseDB(rs, pstmt, conn),你会发现会返回空值。因为应用了该方法,就已经关闭了 rs,没法将数据传递到表现层了。有读者会讲,那为何不在表现层调用该查询方法以后,再去关闭数据库?思路是正确的,但是考虑到封装性和代码可读性的层面,这种做法并不可取。读者可以自行去测试一下。查询方法具体含义看注释,源码如下:

```java
public ArrayList <User> dataquery(String sql) {
    ArrayList <User> list = new ArrayList <User>();   //实例化 ArrayList 的对象 list
    conn = getConnection();//连接数据库
    try {
        pstmt = conn.prepareStatement(sql);
        rs = pstmt.executeQuery();//执行查询语句,并返回结果集存入 rs
        while (rs.next()) {
            User user = new User();//实例化一个 User 对象,来存一条记录
            user.setUserid(rs.getInt("id"));//将数据依次封装到 user 对象里
            user.setUsername(rs.getString("username"));
            user.setUserpwd(rs.getString("userpwd"));
            user.setUsersex(rs.getString("usersex"));
            user.setUserhobby(rs.getString("userhobby"));
            user.setUserjob(rs.getString("userjob"));
            list.add(user);//将封装好一条记录的 user 对象添加到 list 列表里
        }
    }
    catch (SQLException e) {
        e.printStackTrace();
        System.out.println("检索失败");
    }
    finally
    {
```

```
        releaseDB(rs, pstmt, conn);//关闭链接
    }
    return list;//返回符合条件的所有记录
}
```

5) 用 PrepareStatement 返回符合条件的所有记录数量。具体含义看注释。

```
// ******************* prepareStatement 返回符合条件的记录数量 ***********
public int qureycounts(String sql)
{
    int counts = 0;//定义符合条件的记录数量参数
    conn = getConnection();
    try {
        pstmt = conn.prepareStatement(sql);
        rs = pstmt.executeQuery();//执行统计满足符合条件的记录数的 sql 语句,并将
        //统计结果放到 rs 结果集中,该结果集只有一个字段,显示满足条件的记录数.
        while(rs.next())
        {
            counts = rs.getInt(1);//返回统计结果
        }
    }
    catch (SQLException e) {
        e.printStackTrace();
    }
    finally
    {
        releaseDB(rs, pstmt, conn);
    }
    return counts;
}
```

以上,是对 DBoper 类进行了深入分析,从数据库的连接和关闭,到数据库的具体操作,下面我们可以结合图形化界面,实现对图形化界面中操作的数据进行处理了。比如用户的注册,当用户输入注册信息时,我们可以将注册信息封装到一个 user 对象里,并调用 DBoper 类中的 public boolean pstmtinsert(User user)方法,实现将注册信息插入数据库。

第三步:"图形化界面和事件处理——来分析"。

◆ 主界面 Default.java 文件,演示效果图如 5.26 所示。由主界面可知,该系统主要有添加用户、删除用户、修改用户、查询所有用户等功能。

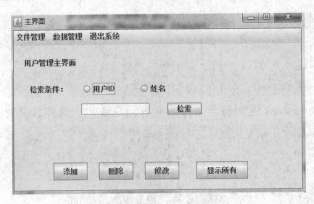

图 5.26 系统主窗体界面

主界面源码如下：

```java
public class Default extends JFrame implements ActionListener {
    private JFrame jf;
    private Container c;
    private JButton btnadd, btndel, btnupdate, btndisplyall, btnqueryone;
    private JRadioButton rbt1;
    private JRadioButton rbt2;
    private JTextField txtquery;
    private JPanel pl;
    private JMenuItem itmadd;
    private JMenuItem itmquery;
    private JMenuItem itmmain;
    private JMenu menuexit;
    private JMenuItem itmclose;
    private JMenuItem itmexit;

    public void defaltDemo() {
        jf = new JFrame("主界面");
        c = jf.getContentPane();
        c.setLayout(null);
        JLabel lbname = new JLabel("用户管理主界面");
        lbname.setBounds(20, 10, 100, 40);
        c.add(lbname);

        btnadd = new JButton("添加");
        btnadd.setBounds(70, 200, 60, 30);
        btnadd.addActionListener(this);
        c.add(btnadd);
```

```java
btndel = new JButton("删除");
btndel.setBounds(150, 200, 60, 30);
btndel.addActionListener(this);
c.add(btndel);

btnupdate = new JButton("修改");
btnupdate.setBounds(230, 200, 60, 30);
btnupdate.addActionListener(this);
c.add(btnupdate);

btndisplyall = new JButton("显示所有");
btndisplyall.setBounds(320, 200, 90, 30);
btndisplyall.addActionListener(this);
c.add(btndisplyall);

JLabel lbquery = new JLabel("检索条件:");
lbquery.setBounds(30, 50, 100, 50);
rbt1 = new JRadioButton("用户ID");
rbt1.setBounds(120, 55, 90, 40);
rbt2 = new JRadioButton("姓名");
rbt2.setBounds(220, 55, 100, 40);
ButtonGroup group = new ButtonGroup();
group.add(rbt2);
group.add(rbt1);
c.add(lbquery);
c.add(rbt2);
c.add(rbt1);
txtquery = new JTextField();
txtquery.setBounds(120, 100, 120, 20);
c.add(txtquery);
btnqueryone = new JButton("检索");
btnqueryone.addActionListener(this);
btnqueryone.setBounds(270, 100, 60, 20);
c.add(btnqueryone);

pl = new JPanel();
pl.setBounds(50, 70, 100, 100);
c.add(pl);
// 菜单界面
```

```java
JMenuBar menubar = new JMenuBar();
c.add(menubar);//新建菜单条,并添加到窗体jf
//新建三个菜单,并把他们添加到菜单条中
JMenu menufile = new JMenu("文件管理");
JMenu menudataM = new JMenu("数据管理");
menuexit = new JMenu("退出系统");
menubar.add(menufile);
menubar.add(menudataM);
menubar.add(menuexit);
//新建三个菜单项添加到"文件管理菜"菜单中
JMenuItem itmopen = new JMenuItem("打开");
JMenuItem itmsave = new JMenuItem("保存");
itmclose = new JMenuItem("关闭");
menufile.add(itmopen);
menufile.add(itmsave);
menufile.add(itmclose);
itmclose.addActionListener(this);
//新建三个菜单项添加到 "数据管理"菜单中
itmadd = new JMenuItem("添加用户");
itmquery = new JMenuItem("查看所有用户");
itmmain = new JMenuItem("新建主界面");
menudataM.add(itmadd);
menudataM.add(itmquery);
menudataM.add(itmmain);
itmadd.addActionListener(this);
itmquery.addActionListener(this);
itmmain.addActionListener(this);
//新建一个菜单项添加到"退出系统"菜单中
itmexit = new JMenuItem("退出");
menuexit.add(itmexit);
itmexit.addActionListener(this);
//让窗体显示出菜单条,这句要放到最后面
jf.setJMenuBar(menubar);
jf.setVisible(true);
jf.setSize(520, 310);
jf.setDefaultCloseOperation(EXIT_ON_CLOSE);//点击关闭时,关闭所有窗体
jf.setDefaultCloseOperation(DISPOSE_ON_CLOSE);
                            //点击关闭按钮时,只关闭当前窗体
}
```

```java
    public static void main(String[] args) {
        Default dm = new Default();
        dm.defaltDemo();
    }
    @Override
    public void actionPerformed(ActionEvent e) {
        String idquerytxt, namequerytxt;
        /** 添加用户信息 **/
            if (e.getSource().equals(btnadd)) {
            addUser adduser = new addUser();
            adduser.adduser();
            jf.dispose();// 关闭当前登陆窗体
            return;
        }
        /** 查询所有用户 **/
        if (e.getSource().equals(btndisplyall)) {
            displaydataquery dy = new displaydataquery();
            String sqlquery = "select * from userme";
            String sqlcount = "select count(*) as count from userme";
            dy.displydata(sqlcount, sqlquery);
            jf.dispose();// 关闭当前登陆窗体
            return;
        }
        /** 按条件进行检索 **/
        if (e.getSource().equals(btnqueryone)) {
            if (rbt1.isSelected()) {
                if (!txtquery.getText().equals("")) {
                    idquerytxt = txtquery.getText();

                    displaydataquery dy = new displaydataquery();
                    String sqlquery = "select * from userme where id = "
                            + Integer.parseInt(idquerytxt);
                    String sqlcount = "select count(*) as count from userme where id = " + Integer.parseInt(idquerytxt);
                    dy.displydata(sqlcount, sqlquery);
                    jf.dispose();// 关闭当前登陆窗体
                    return;
                }
```

```java
            }
            if (rbt2.isSelected()) {
                if (!txtquery.getText().equals("")) {
                    namequerytxt = txtquery.getText().toString();
                    // System.out.println(namequerytxt);
                    displaydataquery dy = new displaydataquery();
                    String sqlquery = "select * from userme where username = '"
                            + namequerytxt + "'";
                    // System.out.println(sqlquery);
                    String sqlcount = "select count( * ) as count from userme where username = '" + namequerytxt + "'";
                    dy.displydata(sqlcount, sqlquery);
                    jf.dispose();// 关闭当前登陆窗体
                    return;
                }
            }
        }
        /** 按条件进行删除 **/
        if (e.getSource().equals(btndel)) {
            if (rbt1.isSelected()) {
                if (!txtquery.getText().equals("")) {
                    idquerytxt = txtquery.getText();
                    String sqlquery = "delete from userme where id = "
                            + Integer.parseInt(idquerytxt);
                    DBoper op = new DBoper();

                    if (op.dataupdate(sqlquery)) {
                        javax.swing.JOptionPane.showMessageDialog(this, "删除成功");
                        new Default().defaltDemo();
                        jf.dispose();// 关闭当前登陆窗体
                        return;
                    } else {
                        javax.swing.JOptionPane.showMessageDialog(this, "删除失败");
                        return;
                    }
                }
            }
            if (rbt2.isSelected()) {
                if (!txtquery.getText().equals("")) {
```

```java
            namequerytxt = txtquery.getText().toString();
            String sqlquery = "delete from userme where username = '"
                    + namequerytxt + "'";
            DBoper op = new DBoper();

            if (op.dataupdate(sqlquery)) {
                javax.swing.JOptionPane.showMessageDialog(this, "删除成功");
                new Default().defaltDemo();
                jf.dispose();// 关闭当前登陆窗体
                return;
            } else {
                javax.swing.JOptionPane.showMessageDialog(this, "删除失败");
                return;
            }
        }
    }
}
/** 按条件进行修改 **/
if (e.getSource().equals(btnupdate)) {
    int id = 0;
    String name, pwd;
    if (rbt1.isSelected()) {
        if (!txtquery.getText().equals("")) {
            idquerytxt = txtquery.getText();
            String sqlquery = "select * from userme where id = "
                    + Integer.parseInt(idquerytxt);
            DBoper op = new DBoper();
            ArrayList list = op.dataquery(sqlquery);
            User user = new User();
            for (int i = 0; i< list.size(); i++) {
                User user2 = new User();
                user2 = (User) list.get(i);
                id = user2.getUserid();
                user.setUsername(user2.getUsername());
                user.setUserpwd(user2.getUserpwd());
                user.setUsersex(user2.getUsersex());
                user.setUserhobby(user2.getUserhobby());
                user.setUserjob(user2.getUserjob());
            }
```

```java
            updateUser upuser = new updateUser();
            upuser.adduser();
            upuser.update(user);
            upuser.setTxtqueryid(id);
            jf.dispose();// 关闭当前登陆窗体
            return;
        }
    }
}
/** 菜单操作 **/
if (e.getSource().equals(itmadd)) {
    addUser adduser = new addUser();
    adduser.adduser();
    jf.dispose();// 关闭当前登陆窗体
    return;
}
if (e.getSource().equals(itmquery)) {
    displaydataquery dy = new displaydataquery();
    String sqlquery = "select * from userme";
    String sqlcount = "select count(*) as count from userme";
    dy.displydata(sqlcount, sqlquery);
    jf.dispose();// 关闭当前登陆窗体
    return;
}
if (e.getSource().equals(itmmain)) {
    Default dm = new Default();
    dm.defaltDemo();
}
if(e.getSource().equals(itmclose)){
    System.exit(0);
}
if(e.getSource().equals(itmexit)){
    System.exit(0);
}
    }
}
```

分析以上源码,该 Default.java 主要包含三部分内容:定义容器和组件、给容器和组件布局和事件处理。定义容器和组件以及布局是由 public void defaltDemo()方法实现的。布局选用的是绝对布局,事件处理都放在了 public void actionPerformed(ActionEvent e)方法中。依据点击不同的按钮,产生不同的事件处理。defaltDemo()方法中的界面组件产生和布局不再赘述,我们主要讲解一下事件处理部分。

◆ 添加用户

当我们在主窗体上点击"添加"按钮时,事件处理部分源码如下:

```
if (e.getSource().equals(btnadd)) {
    addUser adduser = new addUser();
    adduser.adduser();
    jf.dispose(); // 关闭当前登陆窗体
    return;
}
```

也就是说,当我们点击添加按钮时,我们实例化了另外一个类:addUser 类,并调用了 adduser()方法,同时关闭了当前窗体,进入添加用户的窗体界面。我们运行 addUser 类,窗体界面如图 5.27 所示。

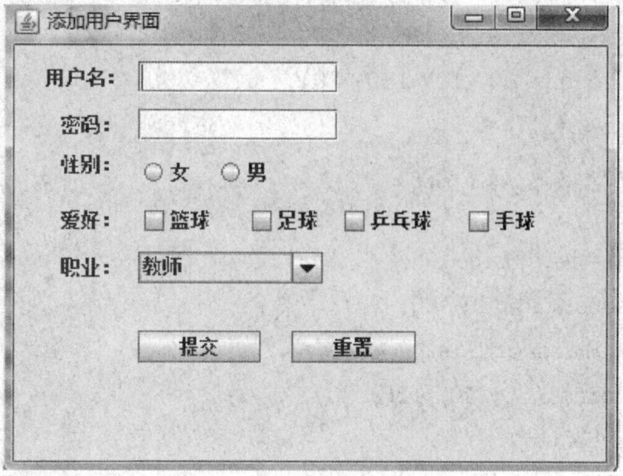

图 5.27 用户添加窗体界面

addUser 类源码如下:

```
package Swing;
import java.awt.Container;
import java.awt.event.ActionEvent;
import java.awt.event.ActionListener;
import javax.swing.*;
import DB.DBoper;
import Dao.User;
public class addUser extends JFrame implements ActionListener {
    private JFrame jf;
    private Container c;
    private JTextField txtname;
    private JPasswordField txtpwd;
    private JRadioButton rbt1;
```

```java
    private JRadioButton rbt2;
    private JCheckBox cb1;
    private JCheckBox cb2;
    private JCheckBox cb3;
    private JCheckBox cb4;
    private JComboBox cbxjob;
    private String usersex;
    private String userhobby = "";
    public void adduser(){
        jf = new JFrame("添加用户界面");
        c = jf.getContentPane();
        c.setLayout(null);
        JLabel lbname = new JLabel("用户名:");
        lbname.setBounds(20, 10, 80, 20);
        txtname = new JTextField();
        txtname.setBounds(80, 10, 130, 20);
        c.add(lbname);
        c.add(txtname);

        JLabel lbpwd = new JLabel("密码:");
        lbpwd.setBounds(30, 40, 80, 20);
        txtpwd = new JPasswordField();
        txtpwd.setBounds(80, 40, 130, 20);
        txtpwd.setEchoChar('*');
        c.add(lbpwd);
        c.add(txtpwd);

        JLabel lbsex = new JLabel("性别:");
        lbsex.setBounds(30, 50, 50, 50);
        rbt1 = new JRadioButton("女");
        rbt1.setBounds(80, 60, 40, 40);
        rbt2 = new JRadioButton("男");
        rbt2.setBounds(130, 60, 40, 40);
        ButtonGroup group = new ButtonGroup();
        group.add(rbt2);
        group.add(rbt1);
        c.add(lbsex);
        c.add(rbt2);
        c.add(rbt1);
```

```java
        JLabel lbhobby = new JLabel("爱好:");
        lbhobby.setBounds(30, 90, 60, 40);
        cb1 = new JCheckBox("篮球");
        cb1.setBounds(80, 90, 60, 40);
        cb2 = new JCheckBox("足球");
        cb2.setBounds(150, 90, 60, 40);
        cb3 = new JCheckBox("乒乓球");
        cb3.setBounds(210, 90, 80, 40);
        cb4 = new JCheckBox("手球");
        cb4.setBounds(290, 90, 60, 40);
        c.add(lbhobby);
        c.add(cb1);
        c.add(cb2);
        c.add(cb3);
        c.add(cb4);

        JLabel lbjob = new JLabel("职业:");
        lbjob.setBounds(30, 120, 40, 40);
        String[] job = {"教师","学生","行政人员"};
        cbxjob = new JComboBox(job);
        cbxjob.setBounds(80,130,120,20);
        c.add(cbxjob);
        c.add(lbjob);

        JButton btnsubmit = new JButton("提交");
        btnsubmit.setBounds(80, 180, 80, 20);
        JButton btnreset = new JButton("重置");
        btnreset.setBounds(180, 180, 80, 20);
        c.add(btnsubmit);
        c.add(btnreset);

        btnsubmit.addActionListener(this);
        btnreset.addActionListener(this);

        jf.setVisible(true);
        jf.setSize(400,300);
        jf.setDefaultCloseOperation(WindowConstants.EXIT_ON_CLOSE);
    }
```

```java
    public static void main(String[] args) {
        addUser win = new addUser();
        win.adduser();
    }
    public void actionPerformed(ActionEvent e) {
        User userDmo = new User();
                                //实例化User类对象userDmo,为封装一条动态数据做准备
        userDmo.setUsername(txtname.getText());
                                        //将录入的用户名set到userDmo对象中
        userDmo.setUserpwd(txtpwd.getText());//将录入的密码set到userDmo对象中
        if(rbt1.isSelected())
            usersex = "女";
        if(rbt2.isSelected())
            usersex = "男";
        userDmo.setUsersex(usersex); //将选择的用户性别set到userDmo对象中

        if(cb1.isSelected())
            userhobby = userhobby + cb1.getText() + "";
        if(cb2.isSelected())
            userhobby = userhobby + cb2.getText() + "";
        if(cb3.isSelected())
            userhobby = userhobby + cb3.getText() + "";
        if(cb4.isSelected())
            userhobby = userhobby + cb4.getText() + "";
        userDmo.setUserhobby(userhobby); //将选择的用户爱好set到userDmo对象中
        userDmo.setUserjob(cbxjob.getSelectedItem().toString());
                                    //将选择的职业set到userDmo对象中
        DBoper db = new DBoper();//实例化数据库操作类,产生db对象
        if (db.pstmtinsert(userDmo)){//调用db的pstmtinsert(userDmo)方法,并传入
//封装好用户注册数据的userDmo对象,插入成功返回真
            javax.swing.JOptionPane.showMessageDialog(this,"成功插入");
                                            //提示插入成功!
            new Default().defaltDemo();//返回主窗体界面
            jf.dispose();//关闭当前登陆窗体
            return;
        }
        else
        {
            javax.swing.JOptionPane.showMessageDialog(this,"插入失败");
            return;
```

```
            }
        }
}
```

窗体界面组件和布局不再赘述，我们还是看事件处理部分。当我们在界面里输入注册信息后，点击提交按钮时，事件处理程序首先实例化一个 User 类的对象 userDmo，并将各个组件的录入信息封装到 userDmo 中，然后，实例化数据库操作类 DBoper 对象 db，调用 db.pstmtinsert(userDmo) 方法，如果返回为真，则说明数据插入成功。这样通过事件处理程序，我们就可以将图像化界面和数据库操作类结合起来，将图像化界面的动态数据，通过数据库操作类里的方法，传递到数据库，当然也可以从数据库取出数据，传递到图形化界面来显示。用户添加部分事件处理程序，详看源码注释。

◆ 删除用户

当我们在主窗体上（Default.java 文件）点击"删除"按钮时，事件处理部分源码如下：

```
/** 按条件进行删除 **/
if (e.getSource().equals(btndel)) {
    if (rbt1.isSelected()) {
        if (!txtquery.getText().equals("")) {
            idquerytxt = txtquery.getText();//获取用户 ID
            String sqlquery = "delete from userme where id = "
                    + Integer.parseInt(idquerytxt);//sql 字符串
            DBoper op = new DBoper();
            if (op.dataupdate(sqlquery)) {
                                        //调用 Dboper 类的 dataupdate 方法，实现删除
                javax.swing.JOptionPane.showMessageDialog(this, "删除成功");
                                                                    //删除成功提示框
                new Default().defaltDemo();//删除成功后，返回主窗体
                jf.dispose();// 关闭当前登陆窗体
                return;
            } else {
                javax.swing.JOptionPane.showMessageDialog(this, "删除失败");
                return;
            }
        }
    }
    if (rbt2.isSelected()) {
        if (!txtquery.getText().equals("")) {
            namequerytxt = txtquery.getText().toString();//获取用户姓名
            String sqlquery = "delete from userme where username = '"
                    + namequerytxt + "'";
            DBoper op = new DBoper();
```

```
            if (op.dataupdate(sqlquery)) {
                javax.swing.JOptionPane.showMessageDialog(this, "删除成功");
                new Default().defaltDemo();
                jf.dispose();// 关闭当前登陆窗体
                return;
            } else {
                javax.swing.JOptionPane.showMessageDialog(this, "删除失败");
                return;
            }
        }
    }
}
```

删除操作如图 5.28 所示。

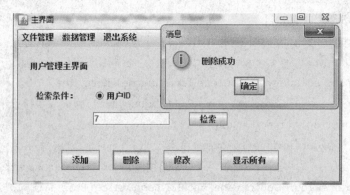

图 5.28　删除用户操作界面

删除条件有两个,一个是按照用户 ID,一个是按照用户姓名。当输入用户 ID,点击删除按钮时,拼出 sql 语句:delete from userme where username='"+ namequerytxt + "';调用数据操作类 DBoper 类里的 dataupdate(sqlquery)方法,执行删除操作,删除完毕以后,给出弹出框提示,删除成功,具体看源码注释。按照姓名删除与按照 ID 删除类似。

◆ 修改用户

当我们在主窗体上(Default.java 文件)点击"修改"按钮时,我们目前以用户 ID 为修改条件来对用户进行修改,以姓名为修改条件的部分由读者尝试完成。以用户 ID 为修改条件事件处理部分源码如下:

```
/** 按条件进行修改【按用户 ID 为条件修改】** /
if (e.getSource().equals(btnupdate)) {
    int id = 0;
    String name, pwd;
    if (rbt1.isSelected()) {
        if (!txtquery.getText().equals("")) {
            idquerytxt = txtquery.getText();// 获取用户 ID
```

```java
            String sqlquery = "select * from userme where id = "
                    + Integer.parseInt(idquerytxt);//拼sql字符串
            DBoper op = new DBoper();
            ArrayList list = op.dataquery(sqlquery);//查出要修改的该用户信息
            User user = new User();//实例化user对象,封装需要修改的用户信息
            for (int i = 0; i < list.size(); i++) {
                User user2 = new User();
                user2 = (User) list.get(i);
                id = user2.getUserid();
                user.setUsername(user2.getUsername());
                user.setUserpwd(user2.getUserpwd());
                user.setUsersex(user2.getUsersex());
                user.setUserhobby(user2.getUserhobby());
                user.setUserjob(user2.getUserjob());
            }
            updateUser upuser = new updateUser();//跳转到用户修改窗体界面
            upuser.adduser();//调用adduser方法,显示修改用户信息主界面
            upuser.update(user);
                        //调用update方法,将部分用户原信息显示在修改界面窗体上
            upuser.setTxtqueryid(id);
            jf.dispose();// 关闭当前登陆窗体
            return;
        }
    }
}
```

用户修改,需要有两个步骤。首先,检索到要修改的用户信息,将用户原信息显示到要修改的界面,如图5.29。当我们输入用户ID 125,然后点击修改按钮时,弹出修改界面,并将ID为125的用户显示修改界面上。具体详情看以上事件处理源码注释。

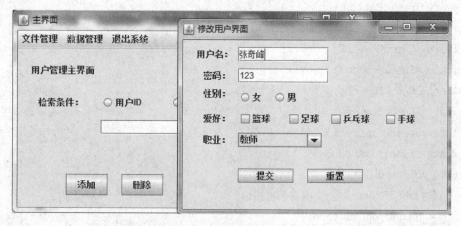

图5.29　修改用户信息显示界面

其次，在修改界面进行修改，点击提交时，将修改后的数据存入数据库，如图 5.30。既然是要将修改页面的新信息存入数据库，我们就要看一下 updateUser.java 文件。

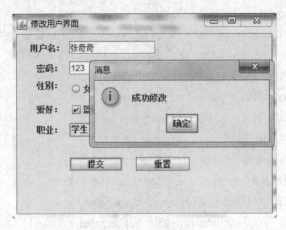

图 5.30 用户修改成功

updateUser.java 主要源码如下：

```java
public class updateUser extends JFrame implements ActionListener {
    private JFrame jf;
    private Container c;
    private JTextField txtname;
    private int txtqueryid;
    public int getTxtqueryid() {
        return txtqueryid;
    }
    public void setTxtqueryid(int id) {
        this.txtqueryid = id;
    }
    private JTextField txtpwd;
    private JRadioButton rbt1;
    private JRadioButton rbt2;
    private JCheckBox cb1;
    private JCheckBox cb2;
    private JCheckBox cb3;
    private JCheckBox cb4;
    private JComboBox cbxjob;
    private String usersex;
    private String userhobby = "";
    public void adduser(){
        jf = new JFrame("修改用户界面");
        c = jf.getContentPane();
```

```java
c.setLayout(null);
JLabel lbname = new JLabel("用户名:");
lbname.setBounds(20, 10, 80, 20);
txtname = new JTextField();
txtname.setBounds(80, 10, 130, 20);
c.add(lbname);
c.add(txtname);
JLabel lbpwd = new JLabel("密码:");
lbpwd.setBounds(30, 40, 80, 20);
txtpwd = new JTextField();
txtpwd.setBounds(80, 40, 130, 20);
c.add(lbpwd);
c.add(txtpwd);
JLabel lbsex = new JLabel("性别:");
lbsex.setBounds(30, 50, 50, 50);
rbt1 = new JRadioButton("女");
rbt1.setBounds(80, 60, 40, 40);
rbt2 = new JRadioButton("男");
rbt2.setBounds(130, 60, 40, 40);
ButtonGroup group = new ButtonGroup();
group.add(rbt2);
group.add(rbt1);
c.add(lbsex);
c.add(rbt2);
c.add(rbt1);
 JLabel lbhobby = new JLabel("爱好:");
 lbhobby.setBounds(30, 90, 60, 40);
 cb1 = new JCheckBox("篮球");
 cb1.setBounds(80, 90, 60, 40);
 cb2 = new JCheckBox("足球");
 cb2.setBounds(150, 90, 60, 40);
 cb3 = new JCheckBox("乒乓球");
 cb3.setBounds(210, 90, 80, 40);
 cb4 = new JCheckBox("手球");
 cb4.setBounds(290, 90, 60, 40);
 c.add(lbhobby);
 c.add(cb1);
 c.add(cb2);
 c.add(cb3);
```

```java
        c.add(cb4);

        JLabel lbjob = new JLabel("职业:");
        lbjob.setBounds(30, 120, 40, 40);
        String[] job = {"教师","学生","行政人员"};
        cbxjob = new JComboBox(job);
        cbxjob.setBounds(80,130,120,20);
        c.add(cbxjob);
        c.add(lbjob);

        JButton btnsubmit = new JButton("提交");
        btnsubmit.setBounds(80, 180, 80, 20);
        JButton btnreset = new JButton("重置");
        btnreset.setBounds(180, 180, 80, 20);
        c.add(btnsubmit);
        c.add(btnreset);

        btnsubmit.addActionListener(this);
        btnreset.addActionListener(this);

        jf.setVisible(true);
        jf.setSize(400,300);
        jf.setDefaultCloseOperation(WindowConstants.EXIT_ON_CLOSE);
    }
    public void update(User user){
        txtname.setText(user.getUsername());
        txtpwd.setText(user.getUserpwd());
    }
    public static void main(String[] args) {
        addUser win = new addUser();
        win.adduser();
    }
    public void actionPerformed(ActionEvent e) {
        User userDmo = new User();//实例化user对象,用来存储修改后的用户信息
        userDmo.setUsername(txtname.getText());//获取修改后的用户名
        userDmo.setUserpwd(txtpwd.getText());//获取修改后的用户密码

        if(rbt1.isSelected())
```

```
        usersex = "女";
    if(rbt2.isSelected())
        usersex = "男";
    userDmo.setUsersex(usersex);  //获取修改后的用户性别
    if(cb1.isSelected())
        userhobby = userhobby + cb1.getText() + "";
    if(cb2.isSelected())
        userhobby = userhobby + cb2.getText() + "";
    if(cb3.isSelected())
        userhobby = userhobby + cb3.getText() + "";
    if(cb4.isSelected())
        userhobby = userhobby + cb4.getText() + "";
    userDmo.setUserhobby(userhobby);  //获取修改后的用户爱好
    userDmo.setUserjob(cbxjob.getSelectedItem().toString());
                                              //获取修改后的用户职业
    DBoper db = new DBoper();  //实例化数据库操作类
    if (db.dataupdateid(userDmo,txtqueryid)){  //调用数据库操作类 dataupdateid
//(userDmo, txtqueryid)方法,实现更新,若返回值为真,说明更新成功
        javax.swing.JOptionPane.showMessageDialog(this,"成功修改");
                                              //弹出修改成功信息
        new Default().defaltDemo();  //返回主窗体
        jf.dispose();  //关闭当前登陆窗体
        return;
    }
    else
    {
        javax.swing.JOptionPane.showMessageDialog(this,"修改失败");
        return;
    }
    }
}
```

由以上源码可以看到,public void adduser( )方法主要显示用户修改界面。public void update(User user)方法是将用户原信息部分显示在修改窗体界面上,目前只显示了用户的用户名和用户密码,读者可以思考如何显示用户的性别和爱好信息。public void actionPerformed(ActionEvent e)方法是事件处理方法,该方法是一个指挥者,它依据图形化界面相应操作来调用数据库操作类或者其他文件里的相应方法,实现图形化界面数据和数据库数据的交互,代码具体含义请看源码注释。

◆ 查看所有用户

当我们在主窗体上(Default.java 文件)点击"显示所有"按钮时,事件处理部分源码如下:

```
/** 查询所有用户 ** /
if (e.getSource().equals(btndisplyall)) {
    displaydataquery dy = new displaydataquery();
    String sqlquery = "select * from userme";
    String sqlcount = "select count( * ) as count from userme";
    dy.displydata(sqlcount, sqlquery);
    jf.dispose();// 关闭当前登陆窗体
    return;
}
```

由以上代码可以看到，当我们点击“查看所有”按钮时，打开了新的窗体类 displaydataquery，并定义了两个 sql 语句，分别是查询所有用户和统计用户总条数。然后调用 displaydataquery 的 displydata(sqlcount, sqlquery)方法，将两个 sql 参数传入。点击“查看所有”按钮时，我们可以看到所有用户信息，如图 5.31 所示。

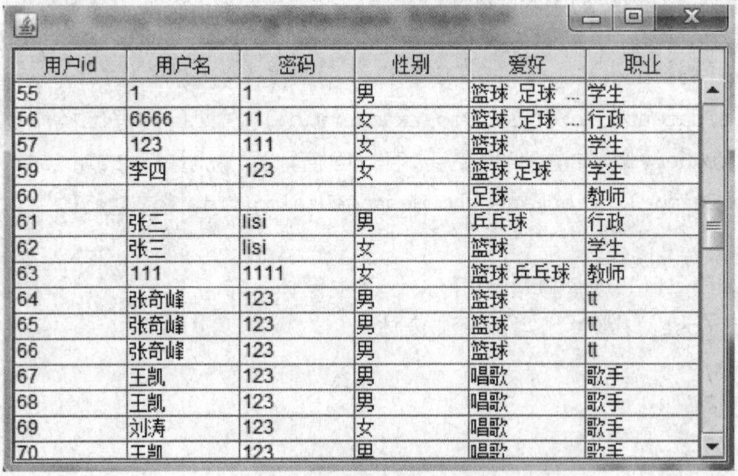

| 用户id | 用户名 | 密码 | 性别 | 爱好 | 职业 |
|---|---|---|---|---|---|
| 55 | 1 | 1 | 男 | 篮球 足球 ... | 学生 |
| 56 | 6666 | 11 | 女 | 篮球 足球 ... | 行政 |
| 57 | 123 | 111 | 女 | 篮球 | 学生 |
| 59 | 李四 | 123 | 女 | 篮球 足球 | 学生 |
| 60 | | | | 足球 | 教师 |
| 61 | 张三 | lisi | 男 | 乒乓球 | 行政 |
| 62 | 张三 | lisi | 女 | 篮球 | 学生 |
| 63 | 111 | 1111 | 女 | 篮球 乒乓球 | 教师 |
| 64 | 张奇峰 | 123 | 男 | 篮球 | tt |
| 65 | 张奇峰 | 123 | 男 | 篮球 | tt |
| 66 | 张奇峰 | 123 | 男 | 篮球 | tt |
| 67 | 王凯 | 123 | 男 | 唱歌 | 歌手 |
| 68 | 王凯 | 123 | 男 | 唱歌 | 歌手 |
| 69 | 刘涛 | 123 | 女 | 唱歌 | 歌手 |
| 70 | 王凯 | 123 | 男 | 唱歌 | 歌手 |

**图 5.31 所有用户信息显示窗体**

下面我们重点看一下窗体类 displaydataquery，其源码如下：

```
public class displaydataquery extends JFrame implements WindowListener{
    private JFrame jf;
    private Container c;
    private ResultSet rs;
    private ArrayList<Integer> strid;
    private ArrayList<String> strname, strpwd, strsex, strhobby, strjob;
    private int counts;
    private JTable table;
    private JScrollPane scrollpane;
    private ArrayList list;
    //计算总共有多少条记录
```

```java
public int countusers(String sql){
    DBoper db = new DBoper();
    counts = db.qureycounts(sql);
    return counts;
}
//获得所有列数据.每一列用 ArrayList 来存
public void displayuser(String sql){
    strid = new ArrayList<Integer>();
                                //实例化一个列表,用来存储所有的用户 id 信息
    strname = new ArrayList<String>();
                                //实例化一个列表,用来存储所有的用户姓名
    strpwd = new ArrayList<String>();
                                //实例化一个列表,用来存储所有的用户密码
    strsex = new ArrayList<String>();
                                //实例化一个列表,用来存储所有的用户性别
    strhobby = new ArrayList<String>();
                                //实例化一个列表,用来存储所有的用户爱好
    strjob = new ArrayList<String>();
                                //实例化一个列表,用来存储所有的用户职业
    DBoper db = new DBoper();
    list = db.dataquery(sql);//从数据库查询所有用户信息
    Iterator it = list.iterator();//实例化一个迭代器,用来迭代所有用户
    while(it.hasNext()) {//循环显示所有用户信息
        User user = new User();
        user = (User) it.next();//循环取出每一条用户记录
        strid.add(user.getUserid());//循环将每一个用户 id 存入 strid 列表
        strname.add(user.getUsername());//循环将每一个用户姓名存入 strname 列表
        strpwd.add(user.getUserpwd());//循环将每一个用户密码存入 strpwd 列表
        strsex.add(user.getUsersex());//循环将每一个用户性别存入 strsex 列表
        strhobby.add(user.getUserhobby());//循环将每一个用户爱好存入 strhobby 列表
        strjob.add(user.getUserjob());//循环将每一个用户职业存入 strjob 列表
    }
}
//将数据表里面的数据存到二维数组里面.
public Object getValueAt(int row, int col) {
    String[][] values = new String[counts][6];
                    //实例化一个二维数组,存储具有 6 个属性的 counts 条记录
    for(row = 1; row<=counts; row++) {
                        //外层循环 counts 次,实现 counts 条记录的读取
```

```java
                    for(col=1; col<=6; col++) {
                                    //内层循环6次,每条记录有6个字段的信息需要读取
                    switch(col) {
            //循环读取各个字段并存入二数组即:将所有用户信息存入二维数组
                    case 1:
            values[row-1][col-1] = strid.get(row-1).toString();
                    break;
                    case 2:
            values[row-1][col-1] = strname.get(row-1);
                    break;
                    case 3:
            values[row-1][col-1]  = strpwd.get(row-1);
                    break;
                    case 4:
            values[row-1][col-1]  = strsex.get(row-1);
                    break;
                    case 5:
            values[row-1][col-1]  = strhobby.get(row-1);
                    break;
                    case 6:
            values[row-1][col-1]  = strjob.get(row-1);
                    break;
                    }
                }
            }
        return values;//返回含有所有用户信息的二维数组
    }
    //Jtable来显示数据
    public void displydata(String sqlcount,String sqlquery){
        jf = new JFrame();
        c = jf.getContentPane();
        String str[] = {"用户id","用户名","密码","性别","爱好","职业"};
                                                    //表头信息一维数组
        displaydataquery t = new displaydataquery();
        t.displayuser(sqlquery);
                //调用displayuser(sqlquery)方法将用户信息存入6个列表里
        int x = t.countusers(sqlcount);
                    //调用countusers(sqlcount)方法统计所用用户数量
        table = new JTable((Object[][]) t.getValueAt(x,5),str);//实例化Jtable,传
//入两个参数,含有所用用户信息的二维数组和含有表头信息的一维数组
```

```java
            table.setBounds(100, 100, 100, 100);//设置Jtable大小和位置
            scrollpane = new JScrollPane(table);//设置Jtable滚动条
            //scrollpane.setBounds(100, 100, 210, 210);
            jf.add(scrollpane);
            jf.addWindowListener(this);
            jf.setVisible(true);
            jf.setSize(400,200);
            jf.setDefaultCloseOperation(DISPOSE_ON_CLOSE);
    }

    public static void main(String[] args) {
        displaydataquery t = new displaydataquery();
        String sqlquery = "select * from userme";

        String sqlcount = "select count( * ) as count from userme";
        t.countusers(sqlcount);
        t.displydata(sqlcount,sqlquery);
    }
    public void windowOpened(WindowEvent e) {

    }
    public void windowClosing(WindowEvent e) {
            new Default().defaltDemo();//当关闭当前窗体时,打开主窗体
    }
    public void windowClosed(WindowEvent e) {
        this.dispose();//当前点击关闭窗体后,让该窗体消失,并打开新的主窗体
        new Default().defaltDemo();
    }
    public void windowIconified(WindowEvent e) {
    }
    public void windowDeiconified(WindowEvent e) {
    }
    public void windowActivated(WindowEvent e) {
    }
    public void windowDeactivated(WindowEvent e) {
    }
}
```

由以上源码可以看到,该类主要有四个方法组成。方法 public int countusers(String sql) 主要是用来统计符合条件的用户记录条数。方法 public void displayuser(String sql)主要是

将符合条件的所有用户取出来,并将每一列单独放置一个列表里。方法 public Object getValueAt(int row, int col)是将每列存储好的列表信息存入一个二维数组里面。方法 public void displydata(String sqlcount,String sqlquery)使用 Jtable 来显示符合条件的所有用户信息。细心的读者会问,为何要先将数据取出来,以每一列存入一个列表的方式存起来,再放置到二维表里,岂不是麻烦? 这里我们之所以这样做,是因为我们显示所有用户的信息时,用的数据结构是 Jtable。这个组件的基本用法里,有两个参数,一个是存有所有信息的二维数组,第二个是需要显示表头信息的一维数组,即: table = new JTable(((Object[ ][ ]) t.getValueAt(x,5),str);当然,读者也可以尝试用其他数据结构来显示用户信息列表。该部分的实现更详细的说明,请看源码注释。值得一提的是,我们这个页面用到了窗口监听事件,当关闭该窗口时,去打开主窗体,所以需要实现 WindowListener 接口里的方法,具体看源码注释。

**第四步,简单总结**

以上是我们对用户管理小系统进行的整体分析和实现。首先,介绍了开发环境所需要的安装文件,以及搭建开发环境的步骤;其次,介绍了数据库的连接,JDBC 数据操作;接下来,结合集合图形化界面和事件处理,实现了该系统的大部分功能,具体主要介绍了用户添加、删除、修改、查询所有用户等功能。系统还有很多需要改进的部分,读者可以继续完善和尝试更复杂的功能。

## 5.5 实验习题

1. 选择题

(1) 下列关于容器的描述中,错误的是( )。
A. 容器是由若干个组件和容器组成的
B. 容器是对图形界面中界面元素的一种管理
C. 容器是一种指定宽和高的矩形范围
D. 容器都是可以独立的窗口

(2) 下列界面元素中,不是容器的是( )。
A. JList          B. JFrame          C. JDialog          D. JPanel

(3) 下列关于实现图形用户界面的描述中,错误的是( )。
A. 放在容器中的组件首先要定义,接着要初始化
B. 放在容器中的多个组件式要进行布局的,默认的布局策略是 FlowLayout
C. 容器中的所有组件都是事件组件,都可产生事件对象
D. 事件处理是由监听者定义的方法来实现的

(4) 下列关于组件类的描述中,错误的是( )。
A. 组件类中包含了文本组件类(JTextComponent)和菜单组件类(JMenuComponent)
B. 标签(JLable)和按钮(JButton)是组件类(JComponent)的子类
C. 面板(JPanel)和窗口(JWindow)是容器类(JContainer)的子类
D. 文本框(JTextField)和文本区(JTextArea)是文本组件类(JTextComponent)的子类

(5) 在对下列语句的解释中,错误的是( )。

but.addActionListener(this);
　A. but 是某种事件对象,如按钮事件对象
　B. this 表示当前容器
　C. ActionListener 是动作事件的监听者
　D. 该语句的功能是将 but 对象注册为 this 对象的监听者
(6) 所有事件类的父类是(　　)。
　A. ActionEvent　　　　　　　　B. AwtEvent
　C. KeyEvent　　　　　　　　　D. MouseEvent
(7) 所有 GUI 标准组件类的父类是(　　)。
　A. JButton　　B. JList　　　C. JComponent　　D. JContainer
(8) 下列各种布局管理器中,Window 类,Dialog 类和 Frame 类的默认布局是(　　)。
　A. FlowLayout　　　　　　　　B. CardLayout
　C. BorderLayout　　　　　　　D. GridLayout
(9) 在下列各种容器中,最简单的无边框的又不能移动和缩放的只能包含在另一种容器中的容器是(　　)。
　A. JWindow　　B. JDialog　　C. JFrame　　　D. JPanel
(10) 下列关于菜单和对话框的描述中,错误的是(　　)。
　A. JFrame 容器是可以容纳菜单组件的容器
　B. 菜单条中可包含若干个菜单,菜单中又包含若干菜单项,菜单项还可包含菜单子项
　C. 对话框与 JFrame 一样都可作为程序的最外层容器
　D. 对话框内不包含有菜单,它是由 JFrame 弹出

2. 编程题
(1) 对 3.3 节项目实训中注册模块进行功能扩展,扩展功能要求如下:
① 将多个用户的注册信息存储到 mysql 数据库表文件里。
② 通过 Swing 的 Jtable 组件,展示数据表里存储的所有用户注册信息列表。
(2) 5.4 综合项目实战项目进行功能扩展,扩展要求如下:
① 实现以用户 ID 和用户姓名为条件的用户查询功能。
② 实现以用户姓名为修改条件的修改功能。
③ 尝试用其他方式来显示用户信息列表。
④ 尝试实现更复杂的菜单功能。

# 实验 6 输入输出流(I/O)

## 6.1 知识点回顾

**1. I/O 流**

变量、数组、引用类型数据都只是临时数据,程序结束后会立即消失。为了长久保存程序处理的数据,需要借助磁盘永久保存数据。Java 使用 I/O 流进行存取操作,其可以将数据保存到文本文件、二进制文件或者 ZIP 压缩包文件。

Java 处理输入输出流的包:java.io.*.

■ 数据流分为输入数据流和输出数据流。

Java 语言定义了许多专门负责各种方式的输入输出,这些类都放在 java.io 包中。其中,所有的输入流都是抽象类 InputStream(字节输入流)或者抽象类 Reader(字符输入流)的子类;而所有的输出流都是抽象类 OutputStream(字节输出流)或者抽象类 Writer(字符输出流)的子类。如图 6.1 所示。

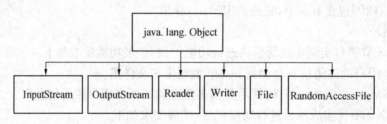

图 6.1 数据流 Java 包类

■ 字节输入流 / 输出流

**InputStream 类是字节输入流的抽象类**,是所有字节输入流的父类,InputStream 类的常用子类的输入流说明如表 6.1 所示,AudioInputStream、ByteArrayInputStream 等九种子类都直接继承于 InputStream。InputStream 类本身不能使用,只能通过继承它的具体类完成某些操作。

它的常用方法如下:

```
public int available()         返回流中可用的字节数
public void close()            关闭流并释放与流相关的系统资源,用户使用完输入流时,调用这个
                               方法
public void mark(int readlimit) 输入流中标志当前位置
public boolean markSupported() 测试流是否支持标志和复位
public abstract int read()     读取输入流中的下一个字节
```

```
public int read(byte[] b)          从输入流中读取字节并存储到缓冲区数组 b 中,返回读
                                   取的字节数,遇到文件结尾返回-1
```

**OutputStream 类是字节输出流的抽象类**,是所有字节输出流的父类,OutputStream 类的常用子类的输出流说明如表 6.1 所示。AudioOutputStream,ByteArrayOutputStream 等九种子类都直接继承于 OutputStream,OutputStream 类本身不能使用,只能通过继承它的具体类完成某些操作。

它的常用方法如下:

```
public void close()   关闭输出流,释放与流相关的系统资源
public void flush()   清洗输出流,使得所有缓冲区的输出字节全部写到输出设备中
public void write(byte[] b)    从特定字节数组 b 将 b 数组长度个字节写入输出流
public void write(byte[] b, int off, int len)
                      从特定字节数组 b 将从 off 开始的 len 个字节写入输出流.
public abstract void write(int b)    向输出流写一个特定字节.
```

表 6.1  字节输入输出流说明

| 字节流描述 | 输入流 | 输出流 |
| --- | --- | --- |
| 音频输入输出流 | AudioInputStream | AudioOutputStream |
| 字节数组输入输出流 | ByteArrayInputStream | ByteArrayOutputStream |
| 文件输入输出流 | FileInputStream | FileOutputStream |
| 过滤器输入输出流 | FilterInputStream | FilterOutputStream |
| 基本输入输出流 | InputStream | OutputStream |
| 对象输入输出流 | ObjectInputStream | ObjectOutputStream |
| 管道输入输出流 | PipedInputStream | PipedOutputStream |
| 顺序输入输出流 | SequenceInputStream | SequenceOutputStream |
| 字符缓冲输入输出流 | StringBufferInputStream | StringBufferOutputStream |

- 字符输入流 / 输出流

Java 中的字符是 Unicode 编码,是双字节的。InputStream 和 OutputStream 是用来处理字节的,并不适合处理字符文本。Java 为字符文本专门提供了一套单独的类 Reader 和 Writer,但是 Reader 和 Writer 并不是 InputStream 和 OutputStream 类的替换者,只是在处理字符串时简化了编程。Reader 和 Writer 是字符输入输出流的抽象类,所有字符输入和输出的实现都是它们的子类。Reader 和 Writer 的子类说明如表 6.2 所示。

表 6.2  Reader/Writer 子类说明

| 字节流描述 | 输入流 | 输出流 |
| --- | --- | --- |
| 将字符数组作为输入输出流 | CharArrayReader | CharArrayWriter |
| 带有默认缓冲的字符输入输出流 | BufferedReader | BufferedRWriter |

(续表)

| 字节流描述 | 输入流 | 输出流 |
| --- | --- | --- |
| 允许过滤字符流 | FilterReader | FilterWriter |
| 从输入输出流读取字节,在将它们转换成字符。 | InputStreamReader | OutputStreamWriter |
| 管道输入输出流 | PipedReader | PipedWriter |
| 读写字符串 | StringReader | StringWriter |

Reader 和 Writer 类中的方法与 InputStream 和 OutputStream 类中的方法类似,读者可以去查看 jdk 文档,不再赘述。

2. File 类

File 类是整个 IO 中与文件本身有关的类,使用 File 类可以创建文件、删除文件等。

➢ 构造方式

```
指定文件路径方式:
File f = newFile(pathName);
为了兼容不同的操作系统,建议 pathName 分隔符采用静态变量方式:
String pathName = "c:" + File.separator + "MainTest.java";
```

➢ 常用方法

```
createNewFile():创建新文件
delete():删除文件
exists():判断文件是否存在
length():字节数
isDirectory():是否为文件目录
isFile():是否为文件
list():返回目录下的文件名字
listFiles():返回目录下文件
mkdir():是否创建了一个目录
renameTo():重命名
```

3. 文件的输入输出流

程序运行期间,大部分数据都在内存中操作,当程序结束或者关闭时,这些数据将消失。如果需要将数据永久保存,可使用文件输入输出流与指定的文件建立连接,将需要的数据永久保存到文件中。

关于流的操作,通常有 3 步:

➢ 建立流对象

➢ 调用流的读/写方法进行数据传输(即输入/输出)

➢ 关闭流

java.io 包中关于文件操作的常用类,如表 6.3 所示。

表 6.3 常用的文件操作类

| 类名 | 用途 |
| --- | --- |
| File | File 创建文件对象，其方法可以实现对文件的操作。 |
| FileInputStream | 继承 InputStream 类，以字节方式实现文件的读取功能。 |
| FileOutputStream | 继承 OutputStream 类，以字节方式实现文件的写入功能。 |
| FileReader | 以字符方式实现文件的读取功能。 |
| FileWriter | 以字符方式实现文件的写入功能。 |
| BufferedInputStream | 带缓存的以字节方式实现文件的读取。 |
| BufferedOutputStream | 带缓存的以字节方式实现文件的写入。 |
| BufferedReader | 继承 Reader 类，带缓存的以字符方式实现文件的读取，可以以行为单位。 |
| BufferedWriter | 继承 Writer 类，带缓存的以字符方式实现文件的写入，可以以行为单位。 |

以上 9 个类的方法和属性，可以查阅 API 文档，详细了解和掌握各个方法的使用。下面以 FileInputStream 类和 BufferedInputStream 类的简单使用为例，将 txt 文本内容读出来。

```java
public static void main(String[] args) {
    File file = new File("c://bcd.txt");
    InputStream io;
    BufferedInputStream bis;
    try {
        io = new FileInputStream(file);
        bis = new BufferedInputStream(io);
        byte [] bytes = new byte[(int)file.length()];
        int temp = 0 ;
        int i = 0 ;
        while((temp = bis.read())  != -1){//判断是否执行到文件最后
            bytes[i] = (byte)temp;
            i++;
        }
        String str = new String(bytes);
        System.out.println(str);
    } catch (FileNotFoundException e) {
        e.printStackTrace();
    } catch (IOException e) {
        e.printStackTrace();
    }
}
```

## 6.2 实验练习

### 6.2.1 实验任务1

● **实验任务**

使用 File 类实现对文件的简单操作:新建文件、查看某一目录下的文件、设置文件权限、过滤文件等。

● **实验要点**
1. 掌握 File 类方法的使用。
2. 定义过滤器。

● **实验分析**
1. 首先要实例化 File 类文件,产生文件对象,然后使用 File 的相关方法,实现对文件的简单操作。
2. 要实现文件过滤,先定义一个过滤器,定义过滤器要实现 FilenameFilter 接口。

● **实验主要代码**
1. 过滤器代码如下:

```java
public class FileSelection implements FilenameFilter {
    String key;
    FileSelection(String key){
        this.key = key;
    }
    public boolean accept(File dir, String name) {
        // TODO Auto-generated method stub
        return name.indexOf(key)! = -1;
    }
}
```

2. 文件简单操作代码如下:

```java
public class fileBaseOperate {
    public static void main(String[] args) throws IOException {
        //新建文件
        File f1 = new File("C:\\B.txt");
        f1.createNewFile();//新建一个 txt 文本
        File f2 = new File("C:\\c.java");
        f2.createNewFile();
```

```java
//判断文件是否 存在
File f3 = new File("C:\\B.txt");
f3.getName();
System.out.println(f3.exists()?"exists":"not exist");

//显示某一路径下所有文件
File path = new File(".");
String[]list = path.list();
for(int i = 0;i<list.length;i++){
    System.out.println(list[i]);
}
File path2 = new File("C://");
String[]list2 = path2.list();
for(int i = 0;i<list2.length;i++){
    System.out.println(list2[i]);
}
//过滤器过滤掉不符合条件的文件(找到同一类别的文件)
System.out.println("++++++++++++++++++++++++++++");
File path3 = new File("C://");
String[] list3 = path3.list(new FileSelection(".java"));
for(int j = 0;j<list3.length;j++){
    System.out.println(list3[j]);
}
//设置文件权限
File testFile = new File("A.java");
testFile.delete();
testFile.createNewFile();
if(testFile.canRead());
System.out.println("A.java 文件可读");
if(testFile.canWrite());
System.out.println("A.java 可写");
testFile.setReadOnly();
if(testFile.canWrite()){
    System.out.println("文件可写");
}
else
{
    System.out.println("A.java 文件不可写!");
}
```

## 6.2.2 实验任务 2

● **实验任务**
1. 以字节方式,模拟从数据库里取出数据,存入 txt 文本。
2. 将 txt 文本内容,读出来。

● **实验要点**
1. 理解并会使用带缓存的文件读写功能。
2. 新建 User 类,理解数据封装性。

● **实验分析**
1. 新建 User 类,该类具有 get 和 set 方法。
2. 实例化多个 User 类的对象,封装数据,并存入 list 里,模拟数据库取出的数据。
3. 将数据通过 FileWriter、BufferedWriter 两个类的方法将数据存入 txt 文本。
4. 通过 FileReader、BufferedReader 两个类的方法将 txt 文本内容以字节方式读出来。

● **实验主要代码**
1. User 类代码如下:

```java
public class User {
    private String id;
    private String username;
    private String userpwd;
    private String useradr;
    private String userhobby;
    private String userinfo;
    private List<User> list;
    private int number;

    public String getUsername() {
        return username;
    }
    public void setUsername(String username) {
        this.username = username;
    }
    public String getUserpwd() {
        return userpwd;
    }
    public void setUserpwd(String userpwd) {
        this.userpwd = userpwd;
    }
    public String getUseradr() {
        return useradr;
```

```
    }
    public void setUseradr(String useradr) {
        this.useradr = useradr;
    }
    public String getUserhobby() {
        return userhobby;
    }
    public void setUserhobby(String userhobby) {
        this.userhobby = userhobby;
    }
    public String getUserinfo() {
        return userinfo;
    }
    public void setUserinfo(String userinfo) {
        this.userinfo = userinfo;
    }
    public String getId() {
        return id;
    }
    public void setId(String id) {
        this.id = id;
    }
    public User(String id, String username, String userpwd, String useradr,
String userhobby, String userinfo) {
        super();
        this.id = id;
        this.username = username;
        this.userpwd = userpwd;
        this.useradr = useradr;
        this.userhobby = userhobby;
        this.userinfo = userinfo;
    }
}
```

2. 带缓存的文件读写代码如下：

```
public class BufferedFile {

    private  List<User> list = new ArrayList<User>();//存放 user 对象的 list
    public List<User> ReturnUsers(){
```

```java
            this.list.add(new User("201101","张三","123","济南","乒乓球","我是应届毕业生"));
            this.list.add(new User("201102","李四","123","南京","乒乓球","我是应届毕业生"));
            this.list.add(new User("201103","王五","123","北京","乒乓球","我是应届毕业生"));
            this.list.add(new User("201104","李二","123","上海","乒乓球","我是应届毕业生"));
            return list;
    }
    /**
     *
     * @将list对象集合以字符方式写入文件.
     */
    public void BufferedWriterFile(List<User> list){
        try {
            File file = new File("C://user.txt");
                if(!file.exists()){
                    file.createNewFile();
                }
            FileWriter fr = new FileWriter(file);
            BufferedWriter buffwriter = new BufferedWriter(fr);
             for(int k = 0;k<list.size();k++){buffwriter.write(list.get(k).getId()+" - "+list.get(k).getUsername()+" - "+list.get(k).getUserpwd()+" - "+list.get(k).getUseradr()+" - "+list.get(k).getUserhobby()+" - "+list.get(k).getUserinfo());
            buffwriter.newLine();
            }
            buffwriter.close();
            fr.close();

        } catch (IOException e) {
            // TODO Auto-generated catch block
            e.printStackTrace();
        }
    }
    /**
     *
     * @将list对象集合以字节方式写入文件.
```

```java
     */
    public void BufferedInputStream(List<User> list){
        
        try {
            File file = new File("C://test.txt");
            if(!file.exists()){
                file.createNewFile();
            }
            FileOutputStream foutwrite = new FileOutputStream(file);
            BufferedOutputStream buffinput = new BufferedOutputStream(foutwrite) ;
            for(int k = 0;k<list.size();k ++ ){
                String s = list.get(k).getId() + " - " + list.get(k).getUsername() + " - " + list.get(k).getUserpwd() + " - " + list.get(k).getUseradr() + " - " + list.get(k).getUserhobby() + " - " + list.get(k).getUserinfo();
                byte[] buf = s.getBytes(); //现将内容放入字节数据,因为 write 方法不能以字符串的形式写入.字节数据是可以的.
                buffinput.write(buf);
            }
            
            buffinput.close();
            foutwrite.close();
            
        } catch (IOException e) {
            // TODO Auto - generated catch block
            e.printStackTrace();
        }
    }
    /**
     *
     * @将 user.txt 文件的内容以字符的方式读出来.
     */
    public void BufferedReaderFile(File file){
        try {
            FileReader in = new FileReader(file);
            BufferedReader bufr = new BufferedReader(in);
            String s = null;
            int i = 0;
            while((s = bufr.readLine())! = null)
```

```java
            {
                i++;
                System.out.println("第"+i+"行:"+s);
            }
            bufr.close();
            in.close();
        } catch (Exception e) {
            // TODO Auto-generated catch block
            e.printStackTrace();
        }
    }
    public static void main(String[] args) {
        BufferedFile bf = new BufferedFile();
        List list = bf.ReturnUsers();
        bf.BufferedWriterFile(list);
        bf.BufferedInputStream(list);
        File file = new File("C://user.txt");
        bf.BufferedReaderFile(file);
    }
}
```

## 6.3 实验习题

1. 选择题

(1) 下列数据流中,属于输入流的一项是(　　)。
A. 从内存流向硬盘的数据流
B. 从键盘流向内存的数据流
C. 从键盘流向显示器的数据流
D. 从网络流向显示器的数据流

(2) Java 语言提供处理不同类型流的类所在的包是(　　)。
A. java.sql　　　　　　　　　　B. java.util
C. java.net　　　　　　　　　　D. java.io

(3) 不属于 java.io 包中的接口的是(　　)。
A. DataInput　　　　　　　　　B. DataOutput
C. DataInputStream　　　　　　D. ObjectInput

(4) 下列程序从标准输入设备读入一个字符,然后再输出到显示器,选择正确的一项填入"//x"处,完成要求的功能(　　)。

```
import java.io.*;
    public class X8_1_4 {
        public static void main(String[] args) {
            char ch;
            try{
                //x
                System.out.println(ch);
            }
            catch(IOException e){
                e.printStackTrace();
            }
        }
    }
```

A. ch = System.in.read();     B. ch = (char)System.in.read();
C. ch = (char)System.in.readln();   D. ch = (int)System.in.read();

（5）下列程序实现了在当前包 dir815 下新建一个目录 subDir815，选择正确的一项填入程序的横线处，使程序符合要求（　）。

```
package dir815;
import java.io.*;
public class X8_1_5 {
public static void main(String[] args){
    char ch;
    try{
        File path = _____;
        if(path.mkdir())
            System.out.println("successful!");
    }
    catch(Exception e){
        e.printStackTrace();
    }
}
}
```

A. new File("subDir815");     B. new File("dir815.subDir815");
C. new File("dir815\subDir815");   D. new File("dir815/subDir815");

6．下列流中哪一个使用了缓冲区技术（　）？
A. BufferedOutputStream     B. FileInputStream
C. DataOutputStream       D. FileReader

（7）能读入字节数据进行 Java 基本数据类型判断过虑的类是（　）。
A. BufferedInputStream     B. FileInputStream

C. DataInputStream  D. FileReader

(8) 使用哪一个类可以实现在文件的任一个位置读写一个记录（　　）?
A. BufferedInputStream  B. RandomAccessFile
C. FileWriter  D. FileReader

(9) 在通常情况下，下列哪个类的对象可以作为 BufferedReader 类构造方法的参数（　　）?
A. PrintStream  B. FileInputStream
C. InputStreamReader  D. FileReader

(10) 若文件是 RandomAccessFile 的实例 f，并且其基本文件长度大于 0，则下面的语句实现的功能是（　　）。
f.seek(f.length()-1);
A. 将文件指针指向文件的第一个字符后面
B. 将文件指针指向文件的最后一个字符前面
C. 将文件指针指向文件的最后一个字符后面
D. 会导致 seek()方法抛出一个 IOException 异常

(11) 下列关于流类和 File 类的说法中错误的一项是（　　）。
A. File 类可以重命名文件  B. File 类可以修改文件内容
C. 流类可以修改文件内容  D. 流类不可以新建目录

(12) 若要删除一个文件，应该使用下列哪个类的实例（　　）?
A. RandomAccessFile  B. File
C. FileOutputStream  D. FileReader

(13) 下列哪一个是 Java 系统的标准输入流对象（　　）?
A. System.out  B. System.in  C. System.exit  D. System.err

(14) Java 系统标准输出对象 System.out 使用的输出流是（　　）。
A. PrintStream  B. PrintWriter
C. DataOutputStream  D. FileReader

2. 编程题

(1) 编写程序，实现读取文件时出现一个表示读取进度的进度条。可以使用 javax.swing 包提供的输入流类 ProgressMonitorInputStream。

(2) 编写程序，使用字符输入、输出流读取文件，将一段文字加密后存入文件，然后在读出来，并将加密前与加密后的文件输出。

**参考文献：**

[1] 明日科技. Java 从入门到精通（第4版）[M]. 北京：清华大学出版社，2012.
[2] Bruce Eckel. Java 编程思想：第4版[M]. 机械工业出版社，2007.
[3] 李刚. 疯狂 Java 讲义. 第3版]M]. 电子工业出版社，2014.

# 实验 7　多线程

## 7.1　知识点回顾

1. 多线程的概念

让系统在同一时间内执行多个程序称为多道程序设计,通常都要涉及进程的概念。一个进程就是一个执行中的程序,都有自己独立的一块内存空间和软硬件组成的系统资源。每一个进程的内部数据和状态都是完全独立的。

线程与进程相似,是一个指令流(一段完成某个特定功能的代码),和进程一样拥有独立的执行控制,由操作系统负责调度,区别在于线程没有独立的存储空间,而是和所属进程中的其他线程共享一个存储空间和系统资源,而线程本身的数据极少。所以系统在同时执行多个线程时,在各个线程之间切换的开销要比进程小很多,正因如此,线程被称为轻负荷进程。

多线程是这样一种机制,它允许在程序中并发执行多个线程(指令流),这意味着一个程序的多行语句可以看上去同时运行。但是这种是在逻辑上"同时",而不是物理上的"同时",除非系统拥有多个 CPU。如果系统只有一个 CPU,由于一块 CPU 同时只能执行一条指令,因此不可能同时执行两个任务。而操作系统为了能提高程序运行的效率,在一个线程空闲时会撤下这个线程,并且会让其他的线程来执行,这种方式叫作线程调度。因为不同线程之间切换的时间非常短,而且非常频繁。因此,从表面看上去就像多个线程同时执行一样,但实际上是交替执行的。

多线程和传统的单线程在程序设计上最大的区别在于,由于各个线程的控制流彼此独立,使得各个线程之间的代码是乱序执行的,将会带来线程调度,同步等问题。

具体到 Java 语言的内存模型,Java 在内存管理上,有一个统一的模型:系统存在一个主内存,Java 中所有变量都储存在主存中,对于所有线程都是共享的。每条线程都有自己的工作内存,工作内存中保存的是主存中某些变量的拷贝,线程对所有变量的操作都是在工作内存中进行,线程之间无法相互直接访问,变量传递均需要通过主存完成。

2. Java 语言多线程实现方法

在 Java 程序中使用多线程要比在 C 或 C++ 中容易得多,这是因为 Java 编程语言提供了语言级的支持。由于 Java 是纯面向对象语言,因此,Java 的线程模型也是面向对象的。Java 通过 Java.lang.Thread 类将线程所必须的功能都封装了起来,并提供了大量的方法来方便控制自己的各个线程,其中最重要的方法是 run()。

```
public class Thread extends Object implements Runnable{}
```

run 方法是线程的执行方法,为了指定自己的线程代码,需要覆盖它。Thread 类还有一个 start 方法负责创建线程,如果线程创建成功,会自动调用 Thread 类的 run 方法。因此,任何继承 Thread 的 Java 类都可以通过 Thread 类的 start 方法来创建线程。

在 Java 的线程模型中除了 Thread 类,还有一个标识某个 Java 类是否可作为线程类的接口 Runnable,这个接口只有一个抽象方法 run,也就是 Java 线程模型的线程执行函数。因此,一个线程类的唯一标准就是这个类是否实现了 Runnable 接口的 run 方法,也就是说,拥有 run( )的类就是线程类。

从上面可以看出,在 Java 中建立线程有两种方法,一种是继承 Thread 类,另一种是实现 Runnable 接口,并通过 Thread 和实现 Runnable 的类来建立线程,其实这两种方法从本质上说是一种方法,即都是通过 Thread 类来建立线程,并运行 run 方法的。但它们的区别是通过继承 Thread 类来建立线程,虽然在实现起来更容易,但由于 Java 不支持多继承,因此,这个线程类如果继承了 Thread,就不能再继承其他的类了。Java 线程模型提供了通过实现 Runnable 接口的方法来建立线程,这样线程类可以在必要的时候继承和业务有关的类,而不是 Thread 类。

任何实现接口 Runnable 的对象都可以作为一个线程的目标对象,类 Thread 本身也实现了接口 Runnable,因此我们可以通过两种方法实现线程体。

方法一:继承 Thread 类

将用户类声明为 Thread 的子类,重写方法 run(),加入线程所要执行的代码,接下来分配并启动该子类的实例即可。下面是一个例子:

```java
class myThread extends Thread {
    public void run() {
    }
}
```

然后,下列代码会创建并启动一个线程:

```java
myThread mt = new myThread();
mt.start();
```

这种方法简单明了,符合大家的习惯,但是,它也有一个很大的缺点,那就是如果用户类已经有一个父类,则无法再继承 Thread 类。

方法二:实现 Runnable 接口

实现 Runnable 接口的类必须使用 Thread 类的实例才能创建线程。通过 Runnable 接口创建线程分为三步:

1. 将实现 Runnable 接口的类实例化。
2. 建立一个 Thread 对象,并将第一步实例化后的对象作为参数传入 Thread 类的构造方法。
3. 通过 Thread 类的 start 方法建立线程。

举例如下:

```java
class myRun implements Runnable {
    public void run() {
    }
}
```

然后,下列代码会创建并启动一个线程:

```java
myRun mr = new myRun();
new Thread(mr).start();
```

综合考虑,使用 Runnable 接口来实现多线程有利于封装。

3. 线程的状态

不管是继承 Thread 类还是实现 Runnable 接口,线程类被执行 new 操作,这时线程就进入了初始状态;当该对象调用了 start()方法,就进入 Runnable(可运行状态);进入 Runnable 后,当该对象被操作系统选中,获得 CPU 时间片就会进入 Running(运行状态)。

进入运行状态后情况比较复杂。通常线程会因为各种原因暂时停止执行,从而退出运行状态,让出 CPU。这里面有设计者通过 Thread 类提供的方法让线程主动让出 CPU 的,也有线程执行完成或者 CPU 时间片到了而自动退出得。这些退出的线程需要执行 Thread 类提供的与之状态对应的方法才能重新进入 Runnable(可运行状态)。可以参看下图 7.1。

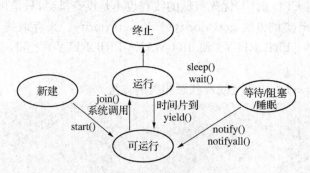

**图 7.1 线程的状态转变**

线程要经历开始、等待(可运行)、运行、挂起(不可运行)和停止几种不同的状态。这些状态都可以通过 Thread 类中的方法进行控制。其中:

(1) 当线程执行 start()方法后,进入可执行状态,但是并不意味着执行,真正进入 CPU 执行,需要等待操作系统的调度。

(2) 当线程调用了 yield()方法,意思是放弃当前获得的 CPU 时间片,回到可运行状态,这意味着该线程与其他线程处于同等竞争状态,这时操作系统将根据调度算法从所有等待的可运行线程中选择一个进入运行状态,也有可能选择的依然是原先的线程。

(3) 当线程的 run()方法或 main()方法结束后,线程就进入终止状态,注意此时线程无法再被调用和执行,但是并不意味着线程不存在,系统依然可以对线程状态和部分资源进行查询,实质上就是还拥有 TCB。

(4) 当线程调用了自身的 sleep()方法或其他线程的 join()方法,就会进入阻塞状态,释放 CPU,但并不释放其他资源。当 sleep()结束或 join()结束后,该线程进入可运行状态,继续等待操作系统分配时间片。

(5) wait() 和 notify() 方法配套使用。wait() 使得线程进入阻塞状态,通常使用无参数的格式,此时线程会释放所占有的所有资源,进入等待队列(与阻塞状态不同),等待状态是不能自动醒来的,必须对应的 notify() 被调用。由于调用 notify() 方法导致解除阻塞的线程是从因调用该对象的 wait() 方法而阻塞的线程中随机选取的,不能确定具体唤醒的是哪一个线程,因此在实际使用时都用 notifyAll()方法唤醒所有的线程。

Object 类的子类每个类实例都有一把锁(可以看作拥有一个取值为 0 或者 1 的变量,1 表示开,0 表示锁),每个 synchronized 方法都必须获得调用该方法的类实例的锁方能执行,否则所属线程阻塞,方法一旦执行,就独占该锁,直到从该方法返回时才将锁释放,此后被阻塞的线程方能获得该锁,重新进入可执行状态(参看下面小节:线程同步问题)。调用 wait() 和 notify()这一对方法的对象上的锁必须为当前线程所拥有,这样 wait()释放所有资源时,才有锁资源可以供释放。因此,这一对方法调用通常都位于 synchronized 方法或块中,该方法或块的上锁对象就是调用这一对方法的对象。若不满足这一条件,则程序虽然仍能编译,但在运行时会出现 IllegalMonitorStateException 异常。

调用 notifyAll() 方法将把因调用该对象的 wait() 方法而阻塞的所有线程一次性全部解除阻塞。但是只有获得锁的那一个线程才能进入可执行状态。

线程的优先级代表该线程的重要程度,当有多个线程同时处于可执行状态并等待获得 CPU 时间时,线程调度系统根据各个线程的优先级来决定给谁分配 CPU 时间,优先级高的线程有更大的机会获得 CPU 时间,优先级低的线程也不是没有机会,只是机会要小一些罢了。

可以调用 Thread 类的方法 getPriority() 和 setPriority()来存取线程的优先级,线程的优先级界于 1(MIN_PRIORITY)和 10(MAX_PRIORITY)之间,缺省是 5(NORM_PRIORITY)。

下面给出了 Thread 类中改变线程状态相关的方法。

```
// 开始线程
public void start( );
public void run( );
// 挂起和唤醒线程
public void resume( );        // 不建议使用
public void suspend( );       // 不建议使用
public static void sleep(long millis);
public static void sleep(long millis, int nanos);
// 终止线程
public void stop( );          // 不建议使用
public void interrupt( );
// 得到线程状态
public boolean isAlive( );
public boolean isInterrupted( );
public static boolean interrupted( );
// join方法
public void join( ) throws InterruptedException;
```

**4. 线程同步问题**

前面提到,多线程和传统的单线程在程序设计上最大的区别在于各个线程之间的代码执行的顺序是不定的,这就引出单线程程序所没有的结果不确定的问题。这部分知识可以参看操作系统。下面只是简单的介绍。

同时执行的进程(线程)之间的关系主要有两种:同步与互斥(这两种关系在操作系统中又被合称为广义同步)。所谓互斥,是指在不同进程(线程)之间的程序片断,当某个进程(线程)运行它的这个程序片段时,其他进程(线程)就不能运行它们之中的这个程序片段,只能等到该进程(线程)运行完这个程序片段后才可以运行。所谓同步,是指在不同进程(线程)之间的程序片断必须严格按照某种规定的顺序来先后运行,这种先后顺序由要完成的特定的任务决定。

简单来说,互斥是两个进程(线程)之间不可以同时运行,必须等待一个进程(线程)运行完毕,另一个才能运行,经常出现在某一具有唯一性和排他性的资源同时只允许一个访问者对其进行访问的情况中。但是互斥无法限制访问者对资源的访问顺序,即访问是无序的。对以上内容不了解的同学请务必参看操作系统的相关章节。

上面说的具有唯一性和排他性的资源,Java 的类变量就是一种。Java 中的变量分为两类:局部变量和类变量。局部变量是指在方法内定义的变量,如在 run()方法中定义的变量。对于这些变量来说,并不存在线程之间共享的问题。因此,它们不需要进行数据同步。类变量是在类中定义的变量,作用域是整个类。这类变量可以被多个线程共享。因此对这类变量进行数据同步。数据同步就是指在同一时间,只能由一个线程来访问被同步的类变量,当前线程访问完这些变量后,其他线程才能继续访问。这里说的访问是指有写操作的访问,如果所有访问类变量的线程都是读操作,一般是不需要数据同步的。

要想解决同步问题,最简单的方法就是使用 synchronized 关键字来使 run 方法同步。这是因为在 Java 语言中,引入了对象互斥锁的概念,来保证共享数据操作的完整性。每个对象都对应于一个可称为"互斥锁"的标记,这个标记用来保证在任一时刻,只能有一个线程访问该对象。关键字 synchronized 来与对象的互斥锁联系。当某个对象用 synchronized 修饰时,表明该对象在任一时刻只能由一个线程访问。加上 synchronized 关键字,就可以使 run 方法同步,也就是说,对于同一个 Java 类的对象实例,run 方法同时只能被一个线程调用,并当前的 run 执行完后,才能被其他的线程调用。即使当前线程执行到了 run 方法中的 yield 方法,也只是暂停了一下。由于其他线程无法执行 run 方法,因此,最终还是会由当前的线程来继续执行,如图 7.2 所示。

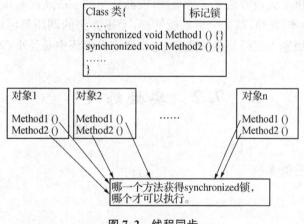

图 7.2 线程同步

synchronized 关键字包括两种用法:synchronized 方法和 synchronized 块。

(1) synchronized 方法:通过在方法声明中加入 synchronized 关键字来声明 synchronized 方法。语法格式:

```
synchronized 返回类型 方法名(参数列表){
    // 其他代码
}
```

synchronized 方法原理是 Java 程序中每个类实例对应一把锁,每个 synchronized 方法都必须获得调用该方法的类实例的锁方能执行,否则所属线程阻塞,方法一旦执行,就独占该锁,直到从该方法返回时才将锁释放,此后被阻塞的线程方能获得该锁,重新进入可执行状态。这种机制确保了同一时刻对于每一个类实例,其所有声明为 synchronized 的成员函数中至多只有一个处于可执行状态(因为至多只有一个能够获得该类实例对应的锁),只要所有可能访问类成员变量的方法均被声明为 synchronized,即可有效避免了类成员变量的访问冲突。

(2) synchronized 块:通过 synchronized 关键字来声明 synchronized 块。语法如下:

```
synchronized(syncObject) {
    //允许访问控制的代码
}
```

若需要将线程类的方法 run() 声明为 synchronized,由于在线程的整个生命期内它一直在运行,因此将导致它对本类任何 synchronized 方法的调用都永远不会成功。类似所有的 synchronized 方法都存在这样的缺陷。解决的方法可以通过将访问类成员变量的代码放到专门的方法中,将其声明为 synchronized,并在主方法中调用来解决。使用 synchronized 块是更好的解决办法。

synchronized 块是一个代码块,其中的代码必须获得对象 syncObject 的锁方能执行,具体机制同前所述。由于可以针对任意代码块,且可任意指定上锁的对象,故灵活性较高。

为了解决对共享存储区的访问冲突,Java 引入了同步机制,但是多个线程对共享资源的访问仅仅依靠同步机制是不够的,因为在任意时刻所要求的资源不一定已经准备好了被访问,反过来,同一时刻准备好了的资源也可能不止一个。为了解决这种情况下的访问控制问题,Java 引入了对阻塞机制的支持。阻塞指的是暂停一个线程的执行以等待某个条件发生(如某资源就绪)。Java 提供了大量方法来支持阻塞,前面介绍的 sleep()、wait()、notify() 都是。谈到阻塞,就不能不谈一谈死锁,略一分析就能发现,上述方法的调用都可能产生死锁。遗憾的是,Java 并不在语言级别上支持死锁的避免,这就要求在编程中必须小心地避免死锁。

## 7.2 实验练习

### 7.2.1 实验任务 1

● 实验任务

编写一个多线程的模拟售票的程序,出售 100 张火车车票,要求座位号码从 1 号到 100 号,用两种方法实现。

● 实验要点

Java 中线程的实现有两种方法:继承 Thread 类和实现 Runnable 接口,如果一个类继承 Thread,就不能继承其他父类,也就不能充分利用父类代码资源。但是这两种方法的区别又并不仅仅是利用接口实现多继承这么简单。

在本实验中火车票的号码不能重复,Runnable 接口法很容易实现这个要求,但是通过继承 Thread 类就不能直接实现,需要在代码中使用静态变量来解决。这个修改,请同学们自己实现。

比较和分析这两种方法可以得出实现 Runnable 接口比继承 Thread 类所具有的另外一个优势:代码和数据独立。这就比较适合同一线程类(模拟售票类)的多个对象线程(独立的售票窗口)去处理同一个资源(车票)这样的问题。

● **实验分析**

模拟售票窗口的问题可以分解成下面的步骤实现:

(1) 定义一个模拟售票类 SimTickets,它的对象就是独立的售票窗口。

(2) 定义一个整形变量 ticketsNum,模拟车票的座位号码 1~100,通过 ticketsNum 逐 1 递减来实现车票座位号的变化。

(3) 通过 Thread 类提供的 getName() 方法取得该线程名,用"线程名"来表示"售票窗口",也可以通过 setName() 方法设置线程名称,但是要注意 setName() 必须在线程启动前执行。

(4) 在屏幕上输出"窗口号"和"座位号",模拟售票的过程:由某窗口售出某座位号车票。

定义一个测试类 Test,通过测试类的 main() 方法启动两个线程或者多个线程,完成模拟售票的过程。

方法一:继承 Thread 类

将用户类声明为 Thread 的子类,重写方法 run(),加入线程所要执行的代码,接下来分配并启动该子类的实例即可。

方法二:实现 Runnable 接口

实现 Runnable 接口的类必须使用 Thread 类的实例才能创建线程。通过 Runnable 接口创建线程分为三步:

(1) 将实现 Runnable 接口的类实例化。

(2) 建立一个 Thread 对象,并将第一步实例化后的对象作为参数传入 Thread 类的构造方法。

(3) 通过 Thread 类的 start 方法建立线程。

● **实验主要代码**

模拟火车票的售票系统,构造一个 Thread 类的子类。

```java
class SimTickets extends Thread
{
    int ticketsNum = 100 ;
    public void run()
    {
        while(true)
        {
            if(ticketsNum>0)
                System.out.println("售出车票----窗口号" + getName());
                System.out.println("----------------座位号" + ticketsNum--);
            else
                System.exit(0);
        }
    }
}
```

}

//测试类:
```
public class Test
{
    public static void main(String[] args)
    {
        SimTickets st1 = new SimTickets();
        SimTickets st2 = new SimTickets();
        st1.start();
        st2.start();
    }
}
```

两次输出结果如图 7.3 所示。

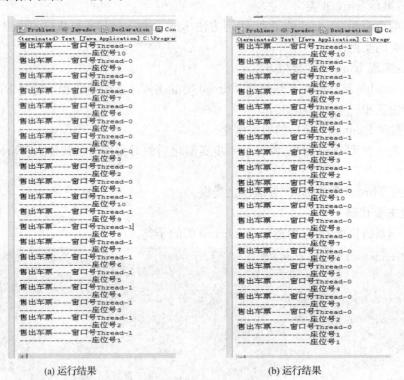

(a) 运行结果　　　　　　　　　(b) 运行结果

图 7.3　继承 Thread 类示例

运行的结果是票被重复出售了,这个错误可以通过为 ticketsNum 添加 static 修饰词来修正。注意运行结果(a)和(b)不一致,表明多线程程序执行的不确定性。

模拟火车票售票系统,利用 Runnable 接口实现。

```
class SimTickets implements Runnable
{
    int ticketsNum = 100 ;
```

```java
    public void run()
    {
        while(true)
        {
            if(ticketsNum>0)
                System.out.println("售出车票--窗口号" + Thread.currentThread().getName());
                System.out.println("------------座位号" + tickets--);
            else
                System.exit(0);
        }
    }
}
//测试类:
public class TestThread
{
    public static void main(String[] args)
    {
        SimTickets st = new SimTickets();      //实例化线程
        Thread  t1 = new Thread (st,"第1售票窗口");
        Thread  t2 = new Thread (st,"第2售票窗口");
        t1.start();
        t2.start();
    }
}
```

两次输出结果如图 7.4 所示。

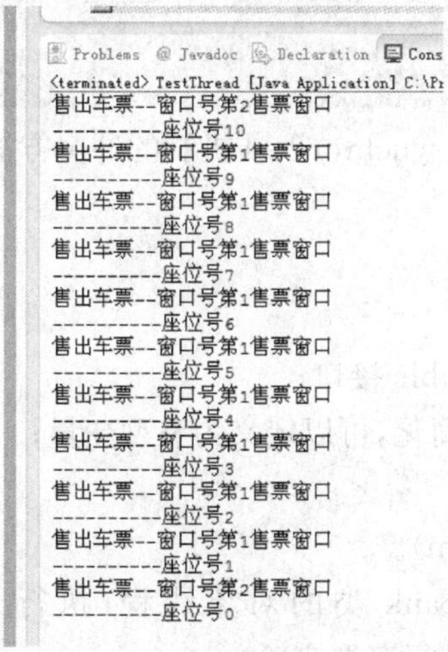

(a) 运行结果 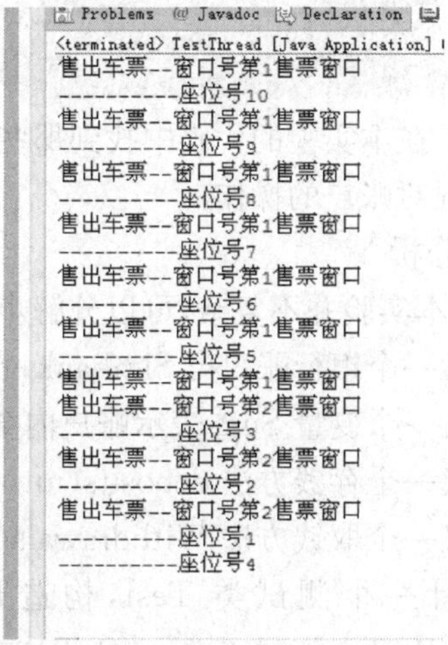　　　　　　　　(b) 运行结果

图 7.4　实现 Runnable 接口示例

注意:如图7.4运行结果(b)所示,本例代码依然存在BUG,虽然这个BUG有较大几率可能不会在实验中显露出来。这个BUG就是:

```
System.out.println("窗口号");
System.out.println("座位号");
```

这两个语句未必是轮流顺序显示的,可能出现连续两个"窗口号"再来连续两个"座位号"的情况。之所以这个BUG未必会暴露,原因很复杂,最大的原因是线程是独立执行的,执行的顺序由系统决定,在系统分配的时间片用完后,即使未能执行完,也必须让出CPU(这部分知识请自行参看操作系统)。所以有可能这两个输出语句会被打断执行,而因为这两个语句很简单,CPU处理很快,所以又有很大概率在一个时间片内执行完成,不会被打断。但是BUG依然存在,如何修正这个BUG,请参看下一个实验。

### 7.2.2 实验任务2

● 实验任务

用多线程编程的方法模拟一个银行账户。可以实现存款和取款操作。

● 实验要点

一个银行账户不管用户同时在几个柜员机上取款和存款,都不会出现数据不一致的现象,例如取出两笔钱,却只减去一笔钱的数目。这意味着当多个线程对同一个账户做操作时,需要实现对账户的互斥访问。

本实验的银行账户互斥访问的问题可以看作是由于同一线程类的多个线程对象共享同一片存储空间而引出的访问冲突问题。Java语言提供了专门的synchronized机制以解决这种互斥冲突,有效避免了同一个数据对象被多个线程同时访问。因为可以通过private关键字来保证数据对象只能被方法访问,所以这套机制是针对方法的,synchronized包括两种用法:synchronized方法和synchronized块。synchronized方法通过在方法声明中加入synchronized关键字来声明。语法格式:

```
synchronized 返回类型 方法名(参数列表){
    // 其他代码
}
```

所以需要在本实验的代码中找到那些必须采用synchronized机制的冲突方法,进行设置,才能正确实现对账户的操作。

● 实验分析

乍一看,本实验并不复杂,可以分解步骤如下:

(1) 定义一个银行账户类SimBank,实现Runnable接口;
(2) 定义一个变量Sum表示账户内金额,为了简化,可以定义为整型变量;
(3) 定义一个存钱方法deposits(int saveSum);
(4) 定义一个取钱方法Withdrawals(int getSum)。

最后设计一个测试类Test,构造几个SimBank类的对象线程,执行deposits(int saveSum)和Withdrawals(int getSum)方法,随机进行存取实验。

这种思路本身不错,问题出在对计算机如何实现账户的存钱和取钱操作考虑不全面。以

存钱方法 deposits(int saveSum)为例，

```
int deposits(int saveSum){
    Sum = Sum + saveSum;
    return Sum;
}
```

这样的代码中就隐藏着 BUG。计算机实现存钱和取钱操作并不想人的大脑一下就可以想出答案，一般来说它需要分几个步骤，先取出存储单元的变量值，放入 CPU 中运算器，运算，存放临时结果，继续运算，得到最终结果，放回存储单元。这个过程具体的步骤和时间的长短，需要考虑到具体的数值、计算机的物理性能、操作系统和 Java 虚拟机的版本等等，如果现实的环境，还需要考虑网络的速度，用户操作的过程。这就导致了在执行的过程中，当前线程很可能被中断，同时另外的线程获得执行，修改了金额等数值，等当前线程恢复执行后，并不能自动获知这个变化，继续操作，导致错误。例如账户有 100 元，现在线程 A 执行取款 100 元操作，程序判断合法，可以执行，此时线程 A 发生中断；线程 B 获得执行，B 线程要取款 100 元，因为 A 线程此时尚未将钱取走，账户有 100 元，所以程序判断合法，可以执行，B 线程完成取款操作；A 线程恢复，继续刚才的执行步骤，接下来就是取款。结果一笔钱被取出两次。

在实验室中，单机环境下，这些现实的环境问题大多数被隐藏起来，存取操作的数值也不大，很可能不出现中断就可以完成执行，这样似乎只要执行一下加减操作就可以完成对账户金额的修改。所以为了能够真正揭示这里面的问题，在实验中把存取钱的步骤里面添加一个临时变量 temp，再利用 Thread.Sleep()方法人为制造一个线程休眠时间。这样数据不一致的问题就暴露出来。

解决这个问题只要在 deposits(int saveSum) 和 withdrawals(int getSum)方法前添加 synchronized 修饰词，引入 synchronized 机制即可。

● 实验主要代码

```
//模拟银行帐户类
class SimBank
{
    private static int Sum = 10000;
    public synchronized static void deposits(int saveSum)
    {
        int temp = Sum;
        temp = temp + saveSum;
        //人为制造线程休眠，时间在 1000 毫秒内随机产生
        try{
            Thread.sleep((int)(1000 * Math.random()));
        }catch(InterruptedException e){
        }
        Sum = temp;
        System.out.println("存入金额:" + saveSum);
```

```java
        System.out.println("当前账户余额:" + Sum);
    }
    public synchronized static void withdrawals(int getSum)
    {
        int temp = Sum;
        temp = temp - getSum;
        //人为制造线程休眠,时间在1000毫秒内随机产生
        try{
          Thread.sleep((int)(1000 * Math.random()));
          }catch(InterruptedException e){
          }
        Sum = temp;
        System.out.println("取出金额:" + getSum);
        System.out.println("当前账户余额:" + Sum);
    }
}
    //模拟用户线程。为了直观,只定义了取钱操作
class Customer extends Thread
{
    public void run()
    {
        //循环10次,每次随机取出1000以内的金额
        for (int i = 1;i<=10 ;i++ )
        {
            SimBank.deposits ((int)(1000 * Math.random()));
        }
    }
}
    //测试类,模拟四个取款机同时对该账户操作
public class Test
{
    public static void main(String[] args)
    {
        Customer c1 = new Customer();
        Customer c2 = new Customer();
        Customer c3 = new Customer();
        Customer c4 = new Customer();
        c1.start();
        c2.start();
```

```
        c3.start();
        c4.start();
    }
}
```

输出结果如图 7.5 所示。

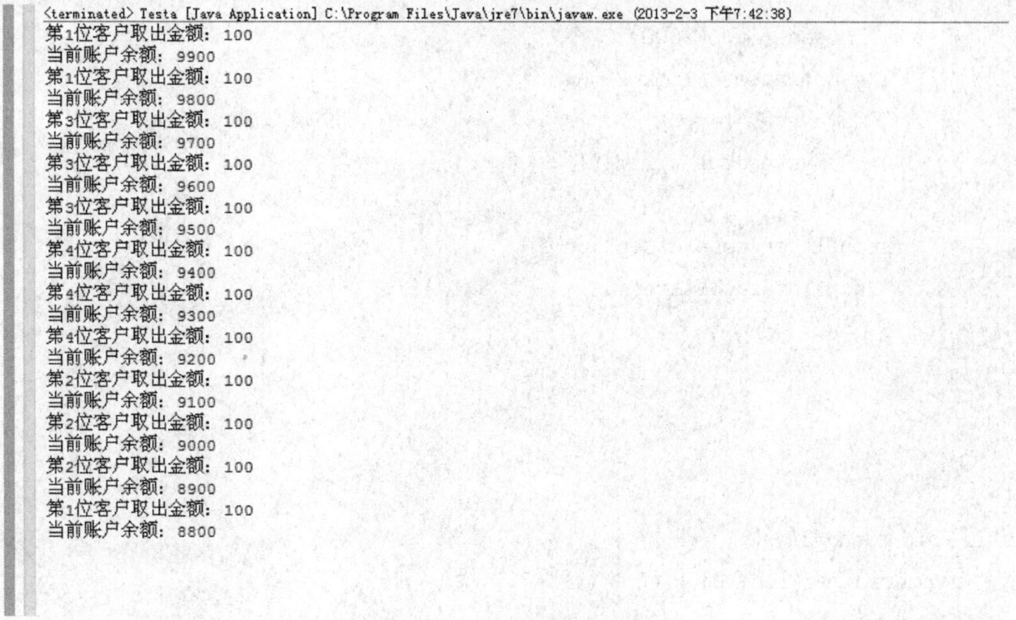

```
<terminated> Testa [Java Application] C:\Program Files\Java\jre7\bin\javaw.exe (2013-2-3 下午7:42:38)
第1位客户取出金额: 100
当前账户余额: 9900
第1位客户取出金额: 100
当前账户余额: 9800
第3位客户取出金额: 100
当前账户余额: 9700
第3位客户取出金额: 100
当前账户余额: 9600
第3位客户取出金额: 100
当前账户余额: 9500
第4位客户取出金额: 100
当前账户余额: 9400
第4位客户取出金额: 100
当前账户余额: 9300
第4位客户取出金额: 100
当前账户余额: 9200
第2位客户取出金额: 100
当前账户余额: 9100
第2位客户取出金额: 100
当前账户余额: 9000
第2位客户取出金额: 100
当前账户余额: 8900
第1位客户取出金额: 100
当前账户余额: 8800
```

图 7.5　多线程运行结果

### 7.2.3　实验任务 3

● 实验任务

用多线程编程的方法模拟一个死锁的线程同步情况。可以实现两个线程互相等待资源出现死锁。

● 实验要点

两个线程在互相调用、互相等待资源造成死锁。死锁也是线程同步中需要避免的情况。通过本实验掌握在同步线程时出现死锁的情况，从而学会避免死锁的发生。

● 实验分析

在类 Deadlocker 中有两个整型成员变量 a,b 和两个 Object 对象 lock_1 和 lock_2 作为类的方法 method1()和 method2()的同步对象。线程 DThread1 和线程 DThread2 都有成员变量 d，分别调用 method1()和 method2()，method1()用 lock_1 作为同步锁，休眠后再用 lock_2 作为同步锁，method2()用 lock_2 作为同步锁，休眠后再用 lock_1 作为同步锁，这个样造成资源互相占用而死锁。

● 实验主要代码

```
class Deadlocker {
    int a,b;
```

```java
private Object lock_1 = new Object();
private Object lock_2 = new Object();

public void method1() {
    synchronized(lock_1) {
        try{
            Thread.sleep(2000);
            synchronized(lock_2) {
                a = 0; b = 0;
                System.out.println("b = " + b);
            }
        }catch(InterruptedException e){
            e.printStackTrace();
        }
    }
}

public void method2() {
    synchronized(lock_2) {
        try{
            Thread.sleep(2000);
            synchronized(lock_1) {
                a = 0; b = 1;
                System.out.println("b = " + b); }
        }catch(InterruptedException e){
            e.printStackTrace();
        }
    }
 }
    public static void main(String []args){
        Deadlocker x = new Deadlocker();
        DThread1 d1 = new DThread1(x);
        DThread2 d2 = new DThread2(x);
         d2.start();
         d1.start();
    }
}
class DThread1 extends Thread{
```

```
        Deadlocker d;
        DThread1(Deadlocker t){
            d = t;
        }
    public void run(){
        d.method1();
    }
}
class DThread2 extends Thread{
    Deadlocker d;
    DThread2(Deadlocker t){
        d = t;
    }
    public void run(){
        d.method2();
    }
}
```

输出结果如图 7.6 所示。

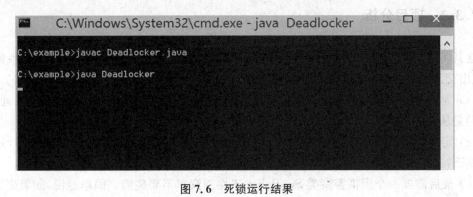

图 7.6 死锁运行结果

## 7.3 项目实战

### 7.3.1 项目描述

结合 Java 图形的知识,做一个汽车过十字路口的动画。要在一个十字路口,交叉的通过几辆汽车,不能出现碰撞事件,如图 7.7 所示。

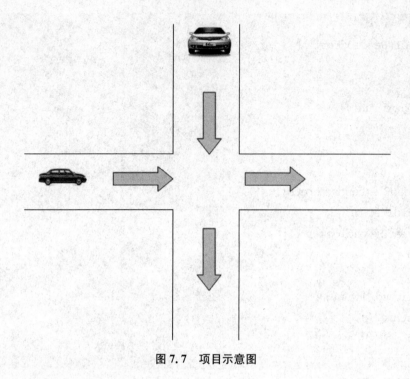

图 7.7 项目示意图

### 7.3.2 项目分析

这其实是对"十字路口"这个资源实现互斥访问。首先可以把项目分解成三部分:做出汽车、做出交叉通过的动画和实现禁止碰撞效果。

(1) 为了突出关键点进行分析和说明,对代码进行简化。同时只有两辆汽车交叉通过,通过方向是从左向右和从上往下,构造汽车类 SimCar。

(2) 定义背景十字路口类 Crossroads,负责绘制黑色的十字路,并加载两辆汽车,还负责定时刷新,这样当汽车的坐标变化时,就呈现动画效果。

(3) 最后需要一个模拟警察类 SimPolice 来控制汽车不要碰撞。简单地说,如果交叉路口有其他汽车,则停止当前汽车线程坐标的变化,即停车。这里稍微烦琐一些,要计算路口的坐标值。需要考虑下一次刷新,是否会碰撞,也就是要用当前的汽车坐标加上汽车速度,看是否和路口的坐标重合,如果没有重合,也就是还没开到路口,不存在碰撞可能,可以继续开车。

### 7.3.3 项目编写

参考代码如下:

```
import Java.applet.Applet;
import Java.awt.*;
public class Crossroads extends Applet implements Runnable
{
    Thread AnimThread = null;
```

```java
//定义两辆汽车
SimCar   LRcar,TBcar;
SimPolice SP;
public void init()
{
    resize(400,400);
    SP = new SimPolice();
    LRcar = new SimCar(SP,SimCar.leftToRight,16);
    TBcar = new SimCar(SP,SimCar.topToBotton,17);
}
public void start()
{
    if (AnimThread = = null)
    {
        AnimThread = new Thread(this);
        AnimThread.start();
        if (LRcar! = null && TBcar! = null)
        {
            LRcar.start();
            TBcar.start();
        }
    }
    else
      {
        AnimThread.resume();    //恢复线程运行
        if (LRcar! = null){LRcar.resume(); }
        if (TBcar! = null){TBcar.resume(); }
      }
}
//挂起线程
public void stop()
{
    AnimThread.suspend();    //挂起线程
    if (LRcar! = null){LRcar.suspend(); }
    if (TBcar! = null){TBcar.suspend(); }
}
public void run()
{
    while(true)
```

```java
            {
                //间隔50毫秒刷新
                try{Thread.sleep(50);}
                catch(InterruptedException e){}
                repaint();
            }
        }
    public void paint(Graphics g)
    {
        //绘制道路
        Color saveColor = g.getColor();
        g.setColor(Color.black);
        g.fillRect(0,180,400,40);
        g.fillRect(180,0,40,400);
        //绘制汽车
        LRcar.drawCar(g);
        TBcar.drawCar(g);
    }
    public void update(Graphics g)
    {
        if(!isValid())
            {
                paint(g);
                return;
            }
        LRcar.drawCar(g);
        TBcar.drawCar(g);
    }
}
class SimCar extends Thread
{
    public int lastPos = -1;
    public int carPos = 0;
    //初始化小车速度
    public int speed = 10;
    //初始化小车的行驶方向
    public int direction = 1;
    public SimPolice SP;
    public final static int leftToRight = 1;
```

```java
    public final static int topToBotton = 2;
    public SimCar(SimPolice SP)
    {
        this(SP,SimCar.leftToRight,10);
    }
    public SimCar(SimPolice SP, int direction, int speed)
    {
        this.SP = SP;
        this.speed = speed;
        this.direction = direction;
    }
    public void run()
    {
        while(true)
        {
            SP.checkAndGo(carPos,speed);
            carPos + = speed;
            if (carPos>= 400)
                { carPos = 0;}
            try{Thread.sleep(200);}
            catch(InterruptedException e){}
        }
    }
//绘制小车
    public void drawCar(Graphics g)
    {
      //方向判断
      if(direction = = SimCar.leftToRight)
        {
           //位置判断
           if(lastPos>= 0)
             {
                //绘制从左往右开的车
                g.setColor(Color.black);
                g.fillRect(0 + lastPos,185,40,32);
             }
             g.setColor(Color.gray);
             g.fillOval(2 + carPos,185,10,10);
             g.fillOval(26 + carPos,185,10,10);
```

```java
                g.fillOval(2 + carPos, 205, 10, 10);
                g.fillOval(26 + carPos, 205, 10, 10);
                g.setColor(Color.green);
                g.fillOval(0 + carPos, 190, 40, 20);
                lastPos = carPos;
            }
            else
            {
                if(lastPos >= 0)
                {
                    //绘制从上往下开的车
                    g.setColor(Color.black);
                    g.fillRect(185, 0 + lastPos, 32, 40);
                }
                g.setColor(Color.gray);
                g.fillOval(185, 2 + carPos, 10, 10);
                g.fillOval(185, 26 + carPos, 10, 10);
                g.fillOval(205, 2 + carPos, 10, 10);
                g.fillOval(205, 26 + carPos, 10, 10);
                g.setColor(Color.yellow);
                g.fillRect(190, 0 + carPos, 20, 40);
                lastPos = carPos;
            }
        }
        public void updateCar(Graphics g)    //更新
        {
            if (lastPos! = carPos)
            {
                drawCar(g);
            }
        }
    }
}
//该类用于 SimCar 线程的控制
class SimPolice
{
    private boolean IntersectionBusy = false;
    //同步化方法
    public synchronizedvoid checkAndGo(int carPos, int speed)
    {
```

```
    //计算坐标
    if(carPos + 40<180 && carPos + 40 + speed>=180 && carPos + speed<=220)
       {
            while(IntersectionBusy)
            {
                try{ wait();}
                catch(InterruptedException e){}
            }

        IntersectionBusy = true;
    }
    if(carPos + speed>220)
    {
      IntersectionBusy = false;
    }
      //线程退出等待状态
      notify();
  }
}
```

## 7.4  实验习题

1. 选择题

(1) Java 语言中提供了一个(   )线程,自动回收动态分配的内存。
   A. 异步　　　　B. 消费者　　　　C. 守护　　　　D. 垃圾收集

(2) 下列说法中错误的一项是(   )。
   A. 线程就是程序
   B. 线程是一个程序的单个执行流
   C. 多线程是指一个程序的多个执行流
   D. 多线程用于实现并发

(3) 下列哪个一个操作不能使线程从等待阻塞状态进入对象阻塞状态(   )。
   A. 等待阻塞状态下的线程被 notify()唤
   B. 等待阻塞状态下的纯种被 interrput()中断
   C. 等待时间到
   D. 等待阻塞状态下的线程调用 wait()方法

(4) 下面的哪一个关键字通常用来对对象的加锁,从而使得对对象的访问是排他的(   )。
   A. sirialize　　　B. transient　　　C. synchronized　　D. static

(5) 哪个方法是实现 Runnable 接口所需的?(   )

A. wait()　　　　B. run()　　　　C. stop()　　　　D. update()

(6) 有三种原因可以导致线程不能运行,它们是(　　)。
A. 等待　　　　　　　　　　B. 阻塞
C. 休眠　　　　　　　　　　D. 挂起及由于 I/O 操作而阻塞

(7) 当(　　)方法终止时,能使线程进入死亡状态。
A. run　　　　B. setPriority　　　　C. yield　　　　D. sleep

(8) 下列说法中错误的一项是(　　)。
A. 一个线程是一个 Thread 类的实例
B. 线程从传递给纯种的 Runnable 实例 run()方法开始执行
C. 线程操作的数据来自 Runnable 实例
D. 新建的线程调用 start()方法就能立即进入运行状态

(9) 用(　　)方法可以改变线程的优先级。
A. run　　　　B. setPriority　　　　C. yield　　　　D. sleep

(10) 线程通过(　　)方法可以休眠一段时间,然后恢复运行。
A. run　　　　B. setPriority　　　　C. yield　　　　D. sleep

2. 编程题

(1) 假设某家银行可接受顾客的汇款,每进行一步汇款,便可计算出汇款总额。现有两名顾客,每人都分 3 次.每次 100 元将钱汇入同一个账户。试编写一个程序,来模拟顾客的汇款操作。

(2) 生产者生产图书,消费者消费图书。由于各个线程是异步运行的,因此无法预计其相对速度,为了使生产者能够不断地生产,可以使用循环缓冲区,保证有足够多的内存区保存更多的产品(生产者——仓库——消费者)。编程模拟实现过程。

# 实验 8  综合项目 1—图像浏览器的实现

## 8.1 项目概述

随着数码相机和手机相机的普及和发展,人们对图像的应用越来越普遍。为了方便查看图像开发了此图像浏览器。

本系统的主要功能模块可以分为用户登录模块、用户注册模块、图像查看模块。用户登录主要通过对登录窗口中输入的用户名和密码与用户信息存放的文件中的信息比对如果一致则登录成功。用户注册是在用户信息文件中添加新的用户名和密码。图像查看模块可对单张和多张图像进行浏览。

## 8.2 项目需求分析

### 8.2.1 功能需求

根据系统的需求,需要创建的内容如下。
(1) 用户登录:根据用户名和密码进行验证,用户名和密码以文本形式保存在文本文件中。
(2) 用户注册:用户名和密码保存的文件中添加新的用户和密码。
(3) 图像查看:图像的选择和查看,查看分为单张查看和多张查看。

## 8.3 总体设计

### 8.3.1 系统界面设计

根据以上的分析,需要设计的登录界面如图 8.1 所示。
(1) 登录界面要求用户输入用户名和密码。
(2) 用户名和密码如果已经存在登录成功否则失败。
(3) 注册部分可以注册新用户,用户名和密码保存在用户文件中。

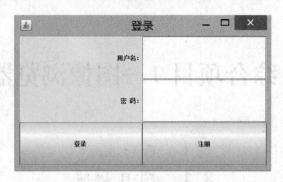

图 8.1 登录界面

根据以上的分析,需要设计的主界面如下 8.2 所示。

(1) 菜单栏:菜单栏包含文件菜单和关于菜单。文件菜单用于选择查看的图像,关于菜单是对系统的简介。

(2) 工具栏:工具栏包含三个按钮,选择了多张图像时分别用来显示上一张、下一张和自动播放。

(3) 显示区:显示区用来显示图像。

(4) 状态栏:状态栏用来显示当前时间和所选的图像信息,如:图形分辨率、大小、位置等。

图 8.2 系统主界面

所以对应有四个类,如表 8.1 所示。

表 8.1 界面设计中的类

| 类 名 | 功能描述 | 所属包 |
| --- | --- | --- |
| PictureShow | 主界面类,程序入口 | swing |
| PlayThread | 浏览图像类 | swing |
| TimeThread | 显示时间类 | swing |
| CreateTray | 系统托盘类 | swing |

根据以上的分析,需要设计的注册界面如下 8.3 所示。

(1) 注册界面要求用户输入用户名和密码。

(2) 用户名和密码如果已经存在则提示已经存在不允许注册。

(3) 注册成功后用户名和密码保存在用户文件中。

图 8.3 用户注册界面

## 8.4 项目文件结构说明

本项目名为:PictureShow 源文件都保存在 src 包中。src 包中下一级包 swing 中存放了所有 java 文件,swing 包中的子包 image 和 user 分别存放了一个图片 lei.jpg 作为系统托盘图标和存放用户名密码的文本文件 user.txt。

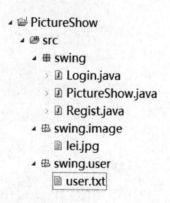

图 8.4 文件结构图

用户名和密码存储在 user.txt 文件中,格式如图 8.5 所示。

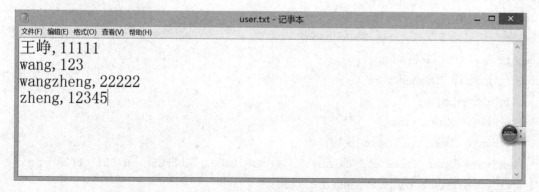

图 8.5 用户文件

## 8.5 主要代码分析

### 8.5.1 登录界面的实现

**关键步骤与代码**

1. 设计登录窗口

(1) 定义一个 Login 类。

(2) 包含的组件：1 个 JFrame 控件，2 个 JLabel 控件，1 个 JTextField 控件，1 个 JPasswordField 控件，2 个 JButton 控件。

(3) 窗口的布局方式：采用边框布局方式，表格布局方式。

运行效果如前面图 8.1 所示，在类中添加如下关键代码：

```java
package swing;
import java.awt.EventQueue;
import javax.swing.JFrame;
import javax.swing.JPanel;
import java.awt.AWTException;
import java.awt.BorderLayout;
import java.awt.GridLayout;
import javax.swing.JLabel;
import javax.swing.JOptionPane;
import javax.swing.JTextField;
import javax.swing.JPasswordField;
import javax.swing.JButton;
import javax.swing.SwingConstants;
import java.awt.event.ActionListener;
import java.awt.event.ActionEvent;
import java.io.BufferedReader;
import java.io.File;
import java.io.FileNotFoundException;
import java.io.FileReader;
import java.io.IOException;
public class Login {
    private JFrame frame;
    private JTextField textField;
    private final JLabel lblNewLabel_1 = new JLabel("\u5BC6\u7801\uFF1A");
    private JPasswordField passwordField;
```

```java
private final JButton loginButton = new JButton("\u767B\u5F55");
private boolean flag = false;
/**
 * Launch the application.
 */
public static void main(String[] args) {
    EventQueue.invokeLater(new Runnable() {
        public void run() {
            try {
                Login window = new Login();
                window.frame.setVisible(true);
            } catch (Exception e) {
                e.printStackTrace();
            }
        }
    });
}

/**
 * Create the application.
 */
public Login() {
    initialize();
}

/**
 * Initialize the contents of the frame.
 */
private void initialize() {
    frame = new JFrame();
    frame.setTitle("\u767B\u5F55");
    frame.setBounds(100, 100, 292, 145);
    frame.setDefaultCloseOperation(JFrame.EXIT_ON_CLOSE);
    
    JPanel panel = new JPanel();
    frame.getContentPane().add(panel, BorderLayout.CENTER);
    panel.setLayout(new GridLayout(0, 2, 0, 0));
    
    JLabel lblNewLabel = new JLabel("\u7528\u6237\u540D\uFF1A");
```

```java
            lblNewLabel.setHorizontalAlignment(SwingConstants.RIGHT);
            panel.add(lblNewLabel);

            textField = new JTextField();
            panel.add(textField);
            textField.setColumns(10);
            lblNewLabel_1.setHorizontalAlignment(SwingConstants.RIGHT);
            panel.add(lblNewLabel_1);

            passwordField = new JPasswordField();
            panel.add(passwordField);
            loginButton.addActionListener(new ActionListener() {
                public void actionPerformed(ActionEvent e) {
                    if(textField.getText().equals("")){
                        JOptionPane.showMessageDialog(null,"用户名不能为空","提示",JOptionPane.ERROR_MESSAGE);
                        return;
                    }
                    //用户文件的读操作
                    File file = new File("src\\swing\\user\\user.txt");
                    FileReader fr = null;
                    BufferedReader br = null;
                    try {
                        fr = new FileReader(file);
                        br = new BufferedReader(fr);
                        String temp;
                        //按行读
                        while((temp = br.readLine())!= null){

                            String t1 = temp.trim();

                            if((textField.getText().trim() + ',' + new String(passwordField.getPassword())).equals(t1)){
                                new PictureShow("图片浏览");
                                flag = true;
                                frame.setVisible(false);
                                break;
                            }
```

```java
                }
                    if(flag==false){
                        JOptionPane.showMessageDialog(null,"用户名不存在,请先注册","提示",JOptionPane.ERROR_MESSAGE);
                        textField.setText("");
                        passwordField.setText("");
                        textField.requestFocus();
                    }
                } catch (FileNotFoundException e1) {
                    // TODO Auto-generated catch block
                    e1.printStackTrace();
                }catch (IOException e2) {
                    // TODO Auto-generated catch block
                    e2.printStackTrace();
                } catch (AWTException e3) {
                    // TODO Auto-generated catch block
                    e3.printStackTrace();
                }finally{
                    
                    try {
                        fr.close();
                        br.close();
                    } catch (IOException e1) {
                        // TODO Auto-generated catch block
                        e1.printStackTrace();
                    }
                }
            }
    });
    panel.add(loginButton);
    
    JButton registButton = new JButton("\u6CE8\u518C");
    registButton.addActionListener(new ActionListener() {
        public void actionPerformed(ActionEvent e) {
            new Regist();
        }
    });
```

```
        panel.add(registButton);
    }
}
```

2. 实现登录功能

匿名实现了 loginButton 和 registerButton 监听,loginButton 监听器用来实现当用户"登录"按钮时从保存用户名和密码的 user.txt 文件中读取内容和文本框中的内容比对。如果用户名和密码都一致则登录成功显示主窗口,否则显示错误提示。而 registerButton 监听器用来显示注册窗口。

### 8.5.2 主窗口的实现

**关键步骤与代码**

1. 设计主窗口

(1) 定义一个 PictureShow 类,继承自 JFrame 框架类。

(2) 窗口中主要包含的组件:1 个 JMenuBar 控件,1 个 JToolBar 控件,2 个 JMenu 控件,3 个 JLabel 控件,3 个 JButton 控件,2 个 JPanel 控件。

(3) 窗口的布局方式:采用边框布局方式。如图 8.2 所示。

在类中添加如下关键代码:

```
package swing;
import java.awt.AWTException;
import java.awt.BorderLayout;
import java.awt.FlowLayout;
import java.awt.Image;
import java.awt.MenuItem;
import java.awt.PopupMenu;
import java.awt.SystemTray;
import java.awt.Toolkit;
import java.awt.TrayIcon;
import java.awt.event.ActionEvent;
import java.awt.event.ActionListener;
import java.io.File;
import java.text.SimpleDateFormat;
import java.util.Date;
import javax.swing.BorderFactory;
import javax.swing.ImageIcon;
import javax.swing.JButton;
import javax.swing.JFileChooser;
import javax.swing.JFrame;
```

```java
import javax.swing.JLabel;
import javax.swing.JMenu;
import javax.swing.JMenuBar;
import javax.swing.JMenuItem;
import javax.swing.JOptionPane;
import javax.swing.JPanel;
import javax.swing.JToolBar;
import javax.swing.UIManager;
import javax.swing.filechooser.FileNameExtensionFilter;

public class PictureShow extends JFrame{
    private static final long serialVersionUID = 1L;
    private JPanel state;//状态面板
    private JLabel statebar;
    private JLabel timebar;
    private JPanel client;//中间显示区
    private JLabel pic;//图片标签
    private TimeThread timethread;
    private PlayThread playthread;
    private JToolBar jtoolbar;
    private File[] files = new File[100];//多张图像选择存放在数组中
    private ImageIcon img;
    private int m = 0;
    private int n = 0;
    public PictureShow(String str) throws AWTException
    {
        super(str);
        //windows 风格
        try {
            //使用 Windows 的界面风格
            UIManager.setLookAndFeel("com.sun.java.swing.plaf.windows.WindowsLookAndFeel");
        } catch (Exception e) {
            e.printStackTrace();
        }

        //菜单栏
        JMenuBar Menubar = new JMenuBar();
        JMenu MenuFile = new JMenu("文件(F)");
        MenuFile.setMnemonic('F');
```

```java
JMenu MenuAbout = new JMenu("关于(H)");
MenuAbout.setMnemonic('H');
Menubar.add(MenuFile);
Menubar.add(MenuAbout);
//菜单项
JMenuItem OpenItem = new JMenuItem("打开图片文件(O)",'O');
OpenItem.setMnemonic('o');

JMenuItem ExitItem = new JMenuItem("退出(X)",'X');
ExitItem.setMnemonic('X');
JMenuItem AboutItem = new JMenuItem("关于(A)",'A');
AboutItem.setMnemonic('A');
MenuFile.add(OpenItem);

MenuFile.addSeparator();
MenuFile.add(ExitItem);
MenuAbout.add(AboutItem);

//菜单项选项
OpenItem.addActionListener(new ActionListener()
{

  public void actionPerformed(ActionEvent e) {
    // TODO 自动生成方法存根
    openfile();
  }
});
ExitItem.addActionListener(new ActionListener()
{
  public void actionPerformed(ActionEvent e){
    System.exit(0);
  }
});

if (!SystemTray.isSupported())
{
    System.out.println("SystemTray is not supported");
    return;
}
```

```java
    //系统托盘
    SystemTray tray = SystemTray.getSystemTray();
        Toolkit toolkit = Toolkit.getDefaultToolkit();
        Image image = toolkit.getImage("src\\swing\\image\\lei.jpg");
        PopupMenu menu = new PopupMenu();
      MenuItem ExitItemTray = new MenuItem("退出");
      ExitItemTray.addActionListener(new ActionListener()
    {
          public void actionPerformed(ActionEvent e){
            System.exit(0);
        }
      });
    menu.add(ExitItemTray);
    TrayIcon icon = new TrayIcon(image,"图片浏览器");
    icon.setImageAutoSize(true);
    icon.setPopupMenu(menu);
    tray.add(icon);
AboutItem.addActionListener(new ActionListener(){
    public void actionPerformed(ActionEvent e) {
      // TODO 自动生成方法存根
      showhelp();
      }


    });
    this.setDefaultCloseOperation(EXIT_ON_CLOSE);

    //设置菜单栏
    super.setJMenuBar(Menubar);


    timebar = new JLabel();
    //时间线程
    timethread = new TimeThread();
    timethread.start();
    //状态栏信息栏
    statebar = new JLabel();
    statebar.setText("未选定");


    client = new JPanel();
    add(client,BorderLayout.CENTER);
```

```java
client.setBorder(BorderFactory.createTitledBorder(""));
pic = new JLabel();
client.add(pic,BorderLayout.CENTER);
pic.setSize(client.getWidth()-200,client.getHeight());
state = new JPanel();
add(state,BorderLayout.SOUTH);//状态栏
state.setBorder(BorderFactory.createTitledBorder(""));
//状态栏两个面板
state.add(timebar);
timebar.setBorder(BorderFactory.createTitledBorder(""));
state.setLayout(new FlowLayout(FlowLayout.LEFT));
state.add(statebar);
jtoolbar = new JToolBar();
JButton jb_next = new JButton("下一个");
jb_next.addActionListener(new ActionListener (){
  public void actionPerformed(ActionEvent e) {
    if(m<0)
       m = 0;
    if(n! = 0){
       img = new ImageIcon(files[m%n].getPath());
       pic.setIcon(img);
       statebar.setText("像素大小:" + img.getIconWidth() + " * " + img.getIconHeight() +"    文件位置:" +
                  files[m%n].getPath().toString()+"    文件大小:" + files[m%n].length()/1024 + "KB"
                  );
       m ++ ;
    }
  }
});
JButton jb_pre = new JButton("上一个");
jb_pre.addActionListener(new ActionListener(){
  public void actionPerformed(ActionEvent arg0) {
     if(m<0)
         m = 0;
     if(n! = 0&&m>=0){
         img = new ImageIcon(files[m%n].getPath());
         pic.setIcon(img);
```

```java
                    statebar.setText("像素大小: " + img.getIconWidth() + " * " + img.getIconHeight() + "    文件位置: " +
                            files[m % n].getPath().toString() + "    文件大小: " + files[m % n].length() / 1024 + "KB"
                    );
            m--;
        }
    });
    JButton jb_play = new JButton("播  放");
    jb_play.addActionListener(new ActionListener(){

        @Override
        public void actionPerformed(ActionEvent arg0) {
            playthread = new PlayThread();
            playthread.start();
        }
    });
    jtoolbar.add(jb_next);
    jtoolbar.add(jb_pre);
    jtoolbar.add(jb_play);
    add(jtoolbar,BorderLayout.NORTH);
    setSize(700,400);//窗口大小
    setLocation(Toolkit.getDefaultToolkit().getScreenSize().width/2 - 350,150);//设置位置
    setVisible(true);//显示
}

//打开图片
public void openfile()
{
    JFileChooser f = new JFileChooser(); // 查找文件
    f.setMultiSelectionEnabled(true);    //设置文件选择器允许选择多个文件
    FileNameExtensionFilter filter = new FileNameExtensionFilter(
            "图片文件(*.jpg,*.gif,*.png)", "jpg", "gif","png");
    f.setFileFilter(filter);
    int x = f.showOpenDialog(this);
    files = f.getSelectedFiles();
```

```java
            n = files.length;

            if(x = = JFileChooser.APPROVE_OPTION){
            img = new ImageIcon(f.getSelectedFile().getPath());
            pic.setIcon(img);
            statebar.setText("像素大小: " + img.getIconWidth() + " * " + img.getIconHeight()
+ "    文件位置: " +
              f.getSelectedFile().getPath().toString() + "文件大小: " + f.getSelectedFile
().length()/1024 + "KB"
            );
            }
            else{
            img = new ImageIcon("src\\swing\\image\\lei.jpg");
            JOptionPane.showMessageDialog(null,"你没有选择图片","请选择图片",JOpt
ionPane.INFORMATION_MESSAGE);
            }
    }

//显示帮助
 public void showhelp()
 {
  JOptionPane.showMessageDialog(this,"这是一款简单的图片浏览器" + "\n" +
    "虽然只能进行图片阅览" + "\n" + "以后改进\n" + "作者:Wallace");
 }
//播放线程
 public class PlayThread extends Thread{
     public void run(){
        for(int i = 0;i<n;i + + )
        {
            img = new ImageIcon(files[i].getPath());
            pic.setIcon(img);
            statebar.setText("像素大小: " + img.getIconWidth() + " * " + img.
getIconHeight() + "    文件位置: " +
                files[i].getPath().toString() + "    文件大小: " + files[i].
length()/1024 + "KB"
                );
            try{
                Thread.sleep(2000);
            }catch(InterruptedException e){}
```

```java
        }
    }
}
//时间线程类
public class TimeThread extends Thread
{
    public void run()
    {
        //不停循环时间刷新
        while(true){
            Date d = new Date();//获取时间
            SimpleDateFormat sdf = new SimpleDateFormat("kk:mm:ss");//转换格式
            timebar.setText(sdf.format(d));
            try{
                Thread.sleep(1000);//1s 执行一次
            }catch(InterruptedException e)
            {
                e.printStackTrace();
            }
        }
    }
}
//托盘类
public class CreateTray
{
    private CreateTray(){}
    private    CreateTray ct = null;
        /*** // ** 创建单实列 * /
        public    CreateTray getInstance()
        {
            //因为使用了判断语句,所以要用 getInstance()方法
            if(ct = = null)
            {
                ct = new CreateTray();
            }
            return ct;
        }
}
```

2. 实现显示功能

通过文件菜单中的打开图片文件子菜单可进行要显示的图像选择,如果选择单张,显示区将直接显示图像;如果按下 Ctrl 键选择多张可通过工具栏中的下一张、上一张切换,也可通过播放按钮自动播放。

### 8.5.3 注册窗体

为了实现增加新用户,注册窗体提供了用户名和密码的添加。用户名信息存放在 swing.user 包中的 user.txt 文件中。

1. 设计注册窗口

(1) 定义一个 Register 类。

(2) 包含的组件:1 个 JFrame 控件,2 个 JLabel 控件,2 个 JTextField 控件,2 个 JButton 控件。

(3) 窗口的布局方式:表格布局方式。

运行效果如前面图 8.3 所示,在类中添加如下关键代码:

```java
package swing;
import java.awt.Image;
import javax.swing.JFrame;
import javax.swing.JPanel;
import java.awt.GridLayout;
import javax.swing.ImageIcon;
import javax.swing.JLabel;
import javax.swing.JOptionPane;
import javax.swing.JTextField;
import javax.swing.SwingConstants;
import javax.swing.border.LineBorder;
import java.awt.Color;
import javax.swing.JButton;
import java.awt.event.ActionListener;
import java.awt.event.ActionEvent;
import java.io.BufferedReader;
import java.io.BufferedWriter;
import java.io.File;
import java.io.FileNotFoundException;
import java.io.FileReader;
import java.io.FileWriter;
import java.io.IOException;
public class Regist {
    private JFrame frame;
    private JTextField textField;
```

```java
    private JTextField textField_1;
    private ImageIcon icon;
    private Image im;
    public Regist() {
        initialize();
    }
    /**
     * Initialize the contents of the frame.
     */
    private void initialize() {
        frame = new JFrame();
        frame.setTitle("\u7528\u6237\u6CE8\u518C");
        frame.setBounds(100, 100, 400, 300);
        frame.setDefaultCloseOperation(JFrame.DISPOSE_ON_CLOSE);
        frame.getContentPane().setLayout(null);
        icon = new ImageIcon("lei.jpg");
        im = icon.getImage();
        frame.setIconImage(im);
        frame.setVisible(true);
        JPanel panel = new JPanel();
        panel.setBorder(new LineBorder(new Color(0, 0, 0), 2));
        panel.setBounds(39, 20, 255, 104);
        frame.getContentPane().add(panel);
        panel.setLayout(new GridLayout(0, 2, 2, 2));
        JLabel lblNewLabel = new JLabel("\u7528\u6237\u540D\uFF1A");
        lblNewLabel.setHorizontalAlignment(SwingConstants.RIGHT);
        panel.add(lblNewLabel);
        textField = new JTextField();
        panel.add(textField);
        textField.setColumns(10);
        JLabel lblNewLabel_1 = new JLabel("\u5BC6\u7801\uFF1A");
        lblNewLabel_1.setHorizontalAlignment(SwingConstants.RIGHT);
        panel.add(lblNewLabel_1);
        textField_1 = new JTextField();
        panel.add(textField_1);
        textField_1.setColumns(10);
        JPanel panel_1 = new JPanel();
        panel_1.setBounds(39, 134, 272, 42);
        frame.getContentPane().add(panel_1);
```

```java
            JButton btnNewButton = new JButton("\u6CE8\u518C");
            btnNewButton.addActionListener(new ActionListener() {
                public void actionPerformed(ActionEvent e) {
                    if(textField.getText().equals("")){
                        JOptionPane.showMessageDialog(null,"用户名不能为空","提示",JOptionPane.ERROR_MESSAGE);
                        return;
                    }
                    File file = new File("src\\swing\\user\\user.txt");
                    FileReader fr = null;
                    BufferedReader br = null;
                    FileWriter fw = null;
                    BufferedWriter bw = null;
                    boolean flag = true;
                    if(!file.exists())
                        try {
                            file.createNewFile();
                        } catch (IOException e1) {
                            // TODO Auto-generated catch block
                            e1.printStackTrace();
                        }
                    try {
                        fr = new FileReader(file);
                        br = new BufferedReader(fr);
                        fw = new FileWriter(file,true);
                        bw = new BufferedWriter(fw);
                        String temp;
                        while((temp = br.readLine())!= null){
                            int n = temp.trim().indexOf(',');
                            String t1 = temp.substring(0, n);
                            String t2 = textField.getText().trim();
                            if(t1.equals(t2)){
                                textField.setText("");
                                JOptionPane.showMessageDialog(null,"用户名已经存在,请换其他名字!","提示",JOptionPane.ERROR_MESSAGE);
                                flag = false;
                                break;
                            }
                        }
                        if(flag){
```

```java
                        fw.write("\r\n" + textField.getText().trim() + ',' + textField_1.getText());
                        fw.flush();
                        JOptionPane.showMessageDialog(null, "注册成功!", "提示", JOptionPane.INFORMATION_MESSAGE);
                    }
                } catch (FileNotFoundException e1) {
                    // TODO Auto-generated catch block
                    e1.printStackTrace();
                } catch (IOException e2) {
                    // TODO Auto-generated catch block
                    e2.printStackTrace();
                }finally{
                  try {
                    fr.close();
                    br.close();
                    br.close();
                    bw.close();
                } catch (IOException e1) {
                    // TODO Auto-generated catch block
                    e1.printStackTrace();
                }
                    
                }
                
            }
        });
        panel_1.add(btnNewButton);
        
        JButton btnNewButton_1 = new JButton("\u53D6\u6D88");
        btnNewButton_1.addActionListener(new ActionListener() {
            public void actionPerformed(ActionEvent arg0) {
                textField.setText("");
                textField_1.setText("");
                textField.requestFocus();
            }
        });
        panel_1.add(btnNewButton_1);
    }
}
```

## 8.6 项目总结

本项目通过对图像浏览器的实现使学习者对Java的UI设计有了初步的了解。通过登录、注册模块的设计使学习者对文件的读写进行了实践操作。图像的显示模块中图像的播放和时间的显示使学习者实践了线程的应用。

**参考文献：**

[1] 武马群.《Java程序设计》.北京:北京工业大学出版社.
[2] 明日科技.《Java从入门到精通(第3版)》.北京:清华大学出版社,2012.
[3] 埃克尔.《JAVA编程思想(第4版)》.机械工业出版社,2007.
[4] 蓝雯飞、周俊陈、淑清,JAVA语言的多态性及其应用研究,计算机系统应用,2005.4.
[5] 吕凤翥、马皓,Java语言程序设计,清华大学出版社,2009.9.
[6] 席国庆,深入体验JAVA项目开发,清华大学出版社,2011.7.
[7] 杨洪雪等,Java开发入门与项目实战,人民邮电出版社,2010.2.
[8] 张思民,Java语言程序设计,清华大学出版社,2007.2.
[9] 王宗亮,Java程序设计任务驱动式实训教程,清华大学出版社,2012.1.

# 实验 9 综合项目 2—图书馆信息管理系统

## 9.1 项目概述

在信息化的今天,对图书管理部门而言,以前单一的手工检索已不能满足人们的要求。为了便于图书资料的管理需要有效的图书信息管理软件,缩短借阅者的等待时间,减轻工作人员的工作量,方便工作人员对它的操作,提高管理的质量和水平,做到高效、智能化管理,同时达到提高图书借阅信息管理效率的目的,采用数据库技术生成的图书馆信息管理系统将会极大地方便借阅者并简化图书馆管理人员和工作人员的劳动,使工作人员从繁忙、复杂的工作进入到一个简单、高效的工作中。本系统利用 Java 语言进行开发,借助 SQL Server 数据库,实现对图书馆信息的管理。主要功能为管理有关读者、图书、借阅和管理者的信息等。

本系统的主要功能模块可以分为读者信息管理模块、图书类别管理模块、图书信息管理模块、新书管理模块、借阅管理模块、系统维护模块。

## 9.2 项目需求分析

### 9.2.1 数据需求

根据系统的需求,需要创建的数据信息如下。

(1) 读者信息:包括读者姓名、年龄、有效证件、证件号码、最大借书量、电话、职业、押金和读者编号等。

(2) 图书类别信息:包括图书类别名称、可借天数和罚款等。

(3) 图书信息:包括图书编号、书名、出版社、出版日期、类别、作者、译者和单价等。

(4) 新书信息:包括订购日期、订购数量、操作员、是否验收、折扣等。

(5) 借阅信息:包括借阅日期、应归还日期、实际归还日期、能否续借、书号和证号等。

(6) 管理者信息:包括用户姓名、性别、年龄、办证日期、联系电话、押金和密码等。

根据这些需求,本系统需要"读者信息"表、"图书类别信息"表、"图书借阅信息"表、"图书订购表"、"图书信息"表和"管理者信息"表。

### 9.2.2 事务需求

(1) 在读者信息管理部分,功能要求如下:

- 可以对读者信息进行添加。
- 可以对读者信息进行修改及删除,并可以查看读者信息。

(2) 在图书类别管理部分,功能要求如下:
- 可以对图书类别进行添加。
- 可以对图书类别进行修改,并可以查看图书类别信息。

(3) 在图书信息管理部分,功能要求如下:
- 可以对图书信息进行添加。
- 可以对图书信息进行修改,并可以查看图书信息。

(4) 在新书管理部分,功能要求如下:
- 可以对新书进行订购。
- 可以对新书进行验收。

(5) 在借阅管理部分,功能要求如下:
- 可以对图书进行借出。
- 可以对图书进行归还。
- 可以对图书进行查询。

(6) 在系统维护部分,功能要求如下:
- 可以对当前的操作员进行密码修改。
- 可以对管理用户进行添加。
- 可以对管理用户进行修改和删除。

## 9.3 项目数据库设计

在 SQL Server 中,先创建 db_library 数据库,然后可以通过 SQL 语句需要生成以下 6 张表。

### 9.3.1 读者信息表(tb_reader)

包括读者姓名、年龄、有效证件、证件号码等,如表 9.1 所示。

表 9.1 读者信息表

| 字段名 | 数据类型 | 大 小 | 可否为空 | 说 明 |
| --- | --- | --- | --- | --- |
| name | varchar | 10 | not null | 读者姓名 |
| sex | varchar | 2 | not null | 读者性别 |
| age | varchar | 4 | not null | 读者年龄 |
| identityCard | varchar | 30 | not null | 读者证件号 |
| date | datetime | 8 | not null | 证件有效日期 |
| maxNum | int | 4 | not null | 最大借书量 |
| tel | varchar | 50 | not null | 读者电话 |

(续表)

| 字段名 | 数据类型 | 大 小 | 可否为空 | 说 明 |
|---|---|---|---|---|
| keepMoney | money | 8 | not null | 读者押金 |
| zj | Int | 4 | not null | 读者证件类型 |
| zy | varchar | 50 | not null | 职业 |
| ISBN | varchar | 13 | not null | 读者编号 |
| **bztime** | datetime | 8 | not null | 办证日期 |

### 9.3.2 图书类别信息表（tb_bookType）

包括图书类别名称、可借天数和罚款等，如表 9.2 所示。

表 9.2 图书类别信息表

| 字段名 | 数据类型 | 大 小 | 可否为空 | 说 明 |
|---|---|---|---|---|
| id | int | 4 | not null | 图书类别编号 |
| typeName | varchar | 20 | not null | 图书类别名称 |
| days | int | 4 | null | 可借天数 |
| **fk** | float | 8 | null | 每天罚款数 |

### 9.3.3 图书信息表（tb_bookInfo）

包括读包括图书编号、书名、出版社等，如表 9.3 所示。

表 9.3 图书信息表

| 字段名 | 数据类型 | 大 小 | 可否为空 | 说 明 |
|---|---|---|---|---|
| ISBN | varchar | 13 | not null | 图书编号 |
| typeId | int | 4 | not null | 图书类别号 |
| bookname | varchar | 40 | not null | 图书名 |
| writer | varchar | 21 | not null | 作者名称 |
| translator | varchar | 30 | null | 译者名称 |
| publisher | varchar | 50 | not null | 出版社名 |
| date | smalldatetime | 4 | not null | 出版日期 |
| **price** | money | 8 | not null | 图书价格 |

### 9.3.4 图书借阅信息表（tb_borrow）

包括借阅日期、应归还日期、实际归还日期、能否续借、书号和证号等，如表 9.4 所示。

表 9.4　图书借阅信息表

| 字段名 | 数据类型 | 大小 | 可否为空 | 说　明 |
|---|---|---|---|---|
| id | int | 4 | not null | 序号 |
| bookISBN | varchar | 13 | null | 图书编号 |
| operatorId | int | 4 | null | 操作员 |
| readerISBN | varchar | 13 | null | 读者编号 |
| isback | int | 4 | not null | 是否已还 |
| borrowDate | datetime | 8 | not null | 借出日期 |
| **backDate** | datetime | 8 | null | 归还日期 |

### 9.3.5　图书订购表(tb_order)

包括订购日期、订购数量、操作员、是否验收、折扣等，如表 9.5 所示。

表 9.5　图书订购表

| 字段名 | 数据类型 | 大小 | 可否为空 | 说　明 |
|---|---|---|---|---|
| ISBN | varchar | 13 | not null | 图书序号 |
| date | datetime | 8 | not null | 订购日期 |
| number | int | 4 | not null | 订购数量 |
| operator | varchar | 6 | not null | 操作员 |
| checkAndAccept | int | 4 | not null | 是否验收 |
| **zk** | float | 8 | not null | 折扣 |

### 9.3.6　管理者信息表(tb_operator)

包括用户姓名、性别、年龄、办证日期、联系电话、押金和密码等，如表 9.6 所示。

表 9.6　管理者信息表

| 字段名 | 数据类型 | 大小 | 可否为空 | 说　明 |
|---|---|---|---|---|
| id | int | 4 | not null | 管理用户编号 |
| name | varchar | 12 | not null | 管理用户名称 |
| sex | varchar | 2 | not null | 性别 |
| age | int | 4 | not null | 年龄 |
| identityCard | varchar | 30 | not null | 证件号 |
| workdate | datetime | 8 | not null | 办证日期 |
| tel | varchar | 50 | not null | 联系电话 |

（续表）

| 字段名 | 数据类型 | 大小 | 可否为空 | 说明 |
|---|---|---|---|---|
| admin | bit | 1 | not null | 是否管理员 |
| **password** | varchar | 10 | not null | 管理用户 |

## 9.4 项目总体设计

### 9.4.1 系统登录界面和主界面设计

根据以上的分析,需要设计的登录界面:
(1) 登录界面要求用户输入用户名和密码。
(2) 用户包括 admin 用户和一般用户。如果是 admin 用户,可以创建其他用户,并且可以修改用户的密码。如果是一般用户,只能修改自己的信息。
(3) 如果用户密码输入正确,可以进入系统主界面。如果输入错误,系统提示出错,可以按下"重置"按钮,继续输入。

图 9.1 登录界面

根据以上的分析,需要设计的主界面如下:
(1) 主界面包括"信息管理"、"新书管理"、"借阅管理"和"系统维护"四个主菜单。
(2) "信息管理"下面分为"读者信息管理"、"图书类别管理"、"图书信息管理"、"退出系统"四个子菜单。"新书管理"下面分为"新书订购"和"验证新书"两个子菜单。"借阅管理"下面分为"图书借阅"、"图书归还"和"图书搜索"三个子菜单。"系统维护"下面分为"更改口令"、"用户管理"两个子菜单。
(3) "读者信息管理"下又分为"读者信息添加"和"读者信息修改与删除"下级菜单,"图书类别管理"下又分为"图书类别添加"和"图书类别修改"下级菜单,"图书信息管理"下又分为"图书信息添加"和"图书信息修改"下级菜单,"用户管理"下又分为"用户添加"和"用户修改与删除"下级菜单。
(4) 在菜单下有 9 个快捷按钮,分别对应"图书信息添加"、"图书信息修改"、"图书类别添加"、"图书借阅"、"新书订购"、"验证新书"、"读者信息添加"、"读者信息修改与删除"、"退出系统"。

图 9.2 系统主界面

所以对应有三个类,如表 9.7 所示。

表 9.7 界面设计中的类

| 类 名 | 功能描述 | 所属包 |
| --- | --- | --- |
| Library | 主界面类,程序入口 | com.lis |
| MenuActions | 设置系统菜单类 | com.lis |
| BookLoginIFrame | 登录界面类 | com.lis.iframe |

### 9.4.2 数据库操作类

对应的数据库操作类有一个,采用纯 Java 驱动实现对数据库的连接,如表 9.8 所示。

表 9.8 数据库操作类

| 类 名 | 功能描述 | 所属包 |
| --- | --- | --- |
| Dao | 数据库连接、对数据库的读写 | com.lis.dao |

### 9.4.3 实体类的设计

对应的实体类如下 9 个类,如表 9.9 所示。

表 9.9 实体类

| 类名 | 功能描述 | 所属包 |
| --- | --- | --- |
| Back | 对表 tb_bookInfo 和表 tb_reader 进行操作 | com.lis.model |
| BookInfo | 对表 tb_bookInfo 进行操作 | com.lis.model |
| BookType | 对表 tb_bookType 进行操作 | com.lis.model |
| Borrow | 对表 tb_borrow 进行操作 | com.lis.model |
| Operater | 对表 tb_operator 进行操作 | com.lis.model |
| Order | 对表 tb_order 进行操作 | com.lis.model |
| OrderAndBookInfo | 对表 tb_bookInfo 和表 tb_order 进行操作 | com.lis.model |
| Reader | 对表 tb_reader 进行操作 | com.lis.model |
| user | 对表 tb_operator 进行操作 | com.lis.model |

### 9.4.4 图书馆信息管理模块类的设计

主要的功能模块所对应的类如下,如表 9.10 所示。

表 9.10 功能模块

| 类名 | 功能描述 | 所属包 |
| --- | --- | --- |
| ReaderAddIFrame | 读者信息添加 | com.lis.iframe |
| ReaderModiAndDelIFrame | 读者信息修改与删除 | com.lis.iframe |
| BookTypeAddIFrame | 图书类别添加 | com.lis.iframe |
| BookTypeModiAndDelIFrame | 图书类别修改 | com.lis.iframe |
| BookAddIFrame | 图书信息添加 | com.lis.iframe |
| BookModiAndDelIFrame | 图书信息修改 | com.lis.iframe |
| newBookOrderIFrame | 新书订购 | com.lis.iframe |
| newBookCheckIFrame | 验证新书 | com.lis.iframe |
| BookBackIFrame | 图书归还 | com.lis.iframe |
| BookBorrowIFrame | 图书借阅 | com.lis.iframe |
| BookSearchIFrame | 图书搜索 | com.lis.iframe |
| UserAddIFrame | 用户添加 | com.lis.iframe |
| ModifyPassword | 密码修改 | com.lis.iframe |
| UserModiAndDelIFrame | 用户修改与删除 | com.lis.iframe |

## 9.5 主要代码分析

### 9.5.1 登录界面的实现

**关键步骤与代码**

1. 设计登录窗口

(1) 定义一个 BookLoginIFrame 类,继承 JFrame 框架类。

(2) 窗口中包含的组件:2 个 JLabel 控件,1 个 JTextField 控件,1 个 JPasswordField 控件,2 个 JButton 控件。

(3) 窗口的布局方式:采用边框布局方式,表格布局方式。

运行效果如前面图 9.1 所示,在类中添加如下关键代码:

```java
public class BookLoginIFrame extends JFrame {
    private JPasswordField password;
    private JTextField username;
    private JButton login;
    private JButton reset;
    private static Operater user;
    public BookLoginIFrame() {
        super();
        final BorderLayout borderLayout = new BorderLayout();
        setDefaultCloseOperation(JFrame.EXIT_ON_CLOSE);
        borderLayout.setVgap(10);
        getContentPane().setLayout(borderLayout);
        setTitle("图书馆管理系统登录");
        setBounds(100, 100, 285, 194);
        final JPanel panel = new JPanel();
        panel.setLayout(new BorderLayout());
        panel.setBorder(new EmptyBorder(0, 0, 0, 0));
        getContentPane().add(panel);
        final JPanel panel_2 = new JPanel();
        final GridLayout gridLayout = new GridLayout(0, 2);
        gridLayout.setHgap(5);
        gridLayout.setVgap(20);
        panel_2.setLayout(gridLayout);
        panel.add(panel_2);
        final JLabel label = new JLabel();
```

```java
label.setHorizontalAlignment(SwingConstants.CENTER);
label.setPreferredSize(new Dimension(0, 0));
label.setMinimumSize(new Dimension(0, 0));
panel_2.add(label);
label.setText("用户名:");
username = new JTextField(20);
username.setPreferredSize(new Dimension(0, 0));
panel_2.add(username);
final JLabel label_1 = new JLabel();
label_1.setHorizontalAlignment(SwingConstants.CENTER);
panel_2.add(label_1);
label_1.setText("密码:");
password = new JPasswordField(20);
password.setDocument(new MyDocument(6));
password.setEchoChar('*');//设置密码框的回显字符
password.addKeyListener(new KeyAdapter() {
    public void keyPressed(final KeyEvent e) {
        if (e.getKeyCode() == 10)
            login.doClick();
    }
});
panel_2.add(password);
final JPanel panel_1 = new JPanel();
panel.add(panel_1, BorderLayout.SOUTH);
login = new JButton();
login.addActionListener(new BookLoginAction());
login.setText("登录");
panel_1.add(login);
reset = new JButton();
reset.addActionListener(new BookResetAction());
reset.setText("重置");
panel_1.add(reset);
final JLabel tupianLabel = new JLabel();
ImageIcon loginIcon = CreatecdIcon.add("login.jpg");
tupianLabel.setIcon(loginIcon);
tupianLabel.setOpaque(true);
tupianLabel.setBackground(Color.GREEN);
tupianLabel.setPreferredSize(new Dimension(260, 60));
panel.add(tupianLabel, BorderLayout.NORTH);
```

```
            setVisible(true);
            setResizable(false);
    }
```

2. 实现登录功能

定义 BookResetAction 监听器和 BookLoginAction 监听器,BookResetAction 监听器用来实现当用户"重置"按钮时清空用户名和密码,而 BookLoginAction 监听器通过 check 方法来查询用户输入的用户名和密码是否正确,如果正确,则进入主界面,否则输出错误信息提示。在类中添加如下关键代码:

```
private class BookResetAction implements ActionListener {
    public void actionPerformed(final ActionEvent e){
        username.setText("");
        password.setText("");
    }
}
class BookLoginAction implements ActionListener {
    public void actionPerformed(final ActionEvent e) {
        user = Dao.check(username.getText(), password.getText());
        if (user.getName() != null) {
            try {
                Library frame = new Library();
                frame.setVisible(true);
                BookLoginIFrame.this.setVisible(false);
            } catch (Exception ex) {
                ex.printStackTrace();
            }
        } else {
            JOptionPane.showMessageDialog(null, "只有管理员才可以登录!");
            username.setText("");
            password.setText("");
        }
    }
}
```

### 9.5.2 主类的实现

**关键步骤与代码**

1. 设计主窗口

(1) 定义一个 Library 类,继承 JFrame 框架类。

(2) 窗口中主要包含的组件:1 个 JMenuBar 控件,1 个 JToolBar 控件,8 个 JMenu 控件,9

个 JButton 控件。

（3）窗口的布局方式：采用边框布局方式。如图 9.2 所示。

在类中添加如下关键代码：

```java
public class Library extends JFrame {
    private static final JDesktopPane DESKTOP_PANE = new JDesktopPane();
    public Library() {
        super();
        setDefaultCloseOperation(WindowConstants.EXIT_ON_CLOSE);
        setLocationByPlatform(true);
        setSize(800, 600);
        setTitle("图书馆管理系统");
        // 调用创建菜单栏的方法
        JMenuBar menuBar = createMenu();
        setJMenuBar(menuBar);
        // 调用创建工具栏的方法
        JToolBar toolBar = createToolBar();
        getContentPane().add(toolBar, BorderLayout.NORTH);
        final JLabel label = new JLabel();
        label.setBounds(0, 0, 0, 0);
        label.setIcon(null);
        DESKTOP_PANE.addComponentListener(new ComponentAdapter() {
            public void componentResized(final ComponentEvent e) {
                Dimension size = e.getComponent().getSize();
                label.setSize(e.getComponent().getSize());
                label.setText("<html><img width=" + size.width + " height="
                        + size.height + " src='"
                        + this.getClass().getResource("/backImg.jpg")
                        + "'></html>");
            }
        });
        DESKTOP_PANE.add(label, new Integer(Integer.MIN_VALUE));
        getContentPane().add(DESKTOP_PANE);
    }
    // 创建工具栏
    private JToolBar createToolBar() {
        JToolBar toolBar = new JToolBar();
        toolBar.setFloatable(false);
        toolBar.setBorder(new BevelBorder(BevelBorder.RAISED));
        JButton bookAddButton = new JButton(MenuActions.BOOK_ADD);
```

```java
ImageIcon icon = new ImageIcon(Library.class.getResource("/bookAddtb.jpg"));
bookAddButton.setIcon(icon);
bookAddButton.setHideActionText(true);
toolBar.add(bookAddButton);
//在工具栏中添加图书修改与删除图标
JButton bookModiAndDelButton = new JButton(MenuActions.BOOK_MODIFY);
ImageIcon bookmodiicon = CreatecdIcon.add("bookModiAndDeltb.jpg");
bookModiAndDelButton.setIcon(bookmodiicon);
bookModiAndDelButton.setHideActionText(true);
toolBar.add(bookModiAndDelButton);
JButton bookTypeAddButton = new JButton(MenuActions.BOOKTYPE_ADD);
ImageIcon bookTypeAddicon = CreatecdIcon.add("bookTypeAddtb.jpg");
bookTypeAddButton.setIcon(bookTypeAddicon);
bookTypeAddButton.setHideActionText(true);
toolBar.add(bookTypeAddButton);
JButton bookBorrowButton = new JButton(MenuActions.BORROW);
ImageIcon bookBorrowicon = CreatecdIcon.add("bookBorrowtb.jpg");
bookBorrowButton.setIcon(bookBorrowicon);
bookBorrowButton.setHideActionText(true);
toolBar.add(bookBorrowButton);
JButton bookOrderButton = new JButton(MenuActions.NEWBOOK_ORDER);
ImageIcon bookOrdericon = CreatecdIcon.add("bookOrdertb.jpg");
bookOrderButton.setIcon(bookOrdericon);
bookOrderButton.setHideActionText(true);
toolBar.add(bookOrderButton);
JButton bookCheckButton = new JButton(MenuActions.NEWBOOK_CHECK_ACCEPT);
ImageIcon bookCheckicon = CreatecdIcon.add("newbookChecktb.jpg");
bookCheckButton.setIcon(bookCheckicon);
bookCheckButton.setHideActionText(true);
toolBar.add(bookCheckButton);
JButton readerAddButton = new JButton(MenuActions.READER_ADD);
ImageIcon readerAddicon = CreatecdIcon.add("readerAddtb.jpg");
readerAddButton.setIcon(readerAddicon);
readerAddButton.setHideActionText(true);
toolBar.add(readerAddButton);
JButton readerModiAndDelButton = new JButton(MenuActions.READER_MODIFY);
ImageIcon readerModiAndDelicon = CreatecdIcon.add("readerModiAndDeltb.jpg");
readerModiAndDelButton.setIcon(readerModiAndDelicon);
readerModiAndDelButton.setHideActionText(true);
```

```java
        toolBar.add(readerModiAndDelButton);
        JButton ExitButton = new JButton(MenuActions.EXIT);
        ImageIcon Exiticon = CreatecdIcon.add("exittb.jpg");
        ExitButton.setIcon(Exiticon);
        ExitButton.setHideActionText(true);
        toolBar.add(ExitButton);
        return toolBar;
    }
    // 创建菜单栏
    private JMenuBar createMenu() {
        JMenuBar menuBar = new JMenuBar();
        // 初始化新书订购管理菜单
        JMenu bookOrderMenu = new JMenu();
        bookOrderMenu.setIcon(CreatecdIcon.add("xsdgcd.jpg"));
        bookOrderMenu.add(MenuActions.NEWBOOK_ORDER);
        bookOrderMenu.add(MenuActions.NEWBOOK_CHECK_ACCEPT);
        // 初始化基础数据维护菜单
        JMenu baseMenu = new JMenu();
        baseMenu.setIcon(CreatecdIcon.add("jcsjcd.jpg"));
        {
            JMenu readerManagerMItem = new JMenu("读者信息管理");
            readerManagerMItem.add(MenuActions.READER_ADD);
            readerManagerMItem.add(MenuActions.READER_MODIFY);
            JMenu bookTypeManageMItem = new JMenu("图书类别管理");
            bookTypeManageMItem.add(MenuActions.BOOKTYPE_ADD);
            bookTypeManageMItem.add(MenuActions.BOOKTYPE_MODIFY);
            JMenu menu = new JMenu("图书信息管理");
            menu.add(MenuActions.BOOK_ADD);
            menu.add(MenuActions.BOOK_MODIFY);
            baseMenu.add(readerManagerMItem);
            baseMenu.add(bookTypeManageMItem);
            baseMenu.add(menu);
            baseMenu.addSeparator();
            baseMenu.add(MenuActions.EXIT);
        }
        // 借阅管理
        JMenu borrowManageMenu = new JMenu();
        borrowManageMenu.setIcon(CreatecdIcon.add("jyglcd.jpg"));
        borrowManageMenu.add(MenuActions.BORROW);
```

```
        borrowManageMenu.add(MenuActions.GIVE_BACK);
        borrowManageMenu.add(MenuActions.BOOK_SEARCH);
        //系统维护
        JMenu sysManageMenu = new JMenu();
        sysManageMenu.setIcon(CreatecdIcon.add("jcwhcd.jpg"));
        JMenu userManageMItem = new JMenu("用户管理");
        userManageMItem.add(MenuActions.USER_ADD);
        userManageMItem.add(MenuActions.USER_MODIFY);
        sysManageMenu.add(MenuActions.MODIFY_PASSWORD);
        sysManageMenu.add(userManageMItem);
        //添加基础数据维护菜单到菜单栏
        menuBar.add(baseMenu);
        //添加新书订购管理菜单到菜单栏
        menuBar.add(bookOrderMenu);
        //添加借阅管理菜单到菜单栏
        menuBar.add(borrowManageMenu);
        menuBar.add(sysManageMenu);
        return menuBar;
    }
}
```

2. 实现主类功能

(1) 在主窗口中要加入系统的入口方法 main 方法，main 方法定义如下：

```
public static void main(String[] args) {
        try {
        Manager.setLookAndFeel(UIManager.getSystemLookAndFeelClassName());
        //登录窗口
        BookLoginIFrame();
        } catch (Exception ex) {
            ex.printStackTrace();
        }
    }
```

(2) 在系统中还要加入菜单操作，所以定义 MenuActions 类，因篇幅有限，只列举了"读者信息添加"菜单和"读者修改与删除"菜单，关键代码如下：

```
public class MenuActions {
    private static Map<String, JInternalFrame> frames; // 子窗体集合
    public static PasswordModiAction MODIFY_PASSWORD;  //修改密码窗体动作
    public static UserModiAction USER_MODIFY;  // 修改用户资料窗体动作
    public static UserAddAction USER_ADD;  // 用户添加窗体动作
    public static BookSearchAction BOOK_SEARCH;  // 图书搜索窗体动作
```

```java
            public static GiveBackAction GIVE_BACK;  // 图书归还窗体动作
            public static BorrowAction BORROW;  // 图书借阅窗体动作
            public static CheckAndAcceptNewBookAction NEWBOOK_CHECK_ACCEPT;
            public static BoodOrderAction NEWBOOK_ORDER;  // 新书定购窗体动作
            public static BookTypeModiAction BOOKTYPE_MODIFY;  //图书类型修改窗
            体动作
            public static BookTypeAddAction BOOKTYPE_ADD;  // 图书类型添加窗体动作
            public static ReaderModiAction READER_MODIFY;  // 读者信息修改窗体动作
            public static ReaderAddAction READER_ADD;  // 读者信息添加窗体动作
            public static BookModiAction BOOK_MODIFY;  // 图书信息修改窗体动作
            public static BookAddAction BOOK_ADD;  // 图书信息添加窗体动作
            public static ExitAction EXIT;  // 系统退出动作
            static {
                    frames = new HashMap<String, JInternalFrame>();
                    MODIFY_PASSWORD = new PasswordModiAction();
                    USER_MODIFY = new UserModiAction();
                    USER_ADD = new UserAddAction();
                    BOOK_SEARCH = new BookSearchAction();
                    GIVE_BACK = new GiveBackAction();
                    BORROW = new BorrowAction();
                    NEWBOOK_CHECK_ACCEPT = new CheckAndAcceptNewBookAction();
                    NEWBOOK_ORDER = new BoodOrderAction();
                    BOOKTYPE_MODIFY = new BookTypeModiAction();
                    BOOKTYPE_ADD = new BookTypeAddAction();
                    READER_MODIFY = new ReaderModiAction();
                    READER_ADD = new ReaderAddAction();
                    BOOK_MODIFY = new BookModiAction();
                    BOOK_ADD = new BookAddAction();
                    EXIT = new ExitAction();
            }
private static class ReaderModiAction extends AbstractAction {
            ReaderModiAction() {
                super("读者修改与删除", null);
                putValue(Action.LONG_DESCRIPTION, "修改和删除读者的基本信息");
                putValue(Action.SHORT_DESCRIPTION, "读者修改与删除");
            }
public void actionPerformed(ActionEvent e) {
    if (!frames.containsKey("读者信息修改与删除")||frames.get("读者信息修改与删
    除").isClosed()) {
```

```java
            ReaderModiAndDelIFrame iframe = new
            ReaderModiAndDelIFrame();
            frames.put("读者信息修改与删除", iframe);
            Library.addIFame(frames.get("读者信息修改与删除"));
        }
    }
}
private static class ReaderAddAction extends AbstractAction {
    ReaderAddAction() {
        super("读者信息添加", null);
        putValue(Action.LONG_DESCRIPTION, "为图书馆添加新的读者会员信息");
        putValue(Action.SHORT_DESCRIPTION, "读者信息添加");
    }
    public void actionPerformed(ActionEvent e) {
        if (!frames.containsKey("读者相关信息添加")||frames.get("读者相关信息添加
        ").isClosed()) {
            ReaderAddIFrame iframe = new ReaderAddIFrame();
            frames.put("读者相关信息添加", iframe);
            Library.addIFame(frames.get("读者相关信息添加"));
        }
    }
}
```

### 9.5.3 实体类的实现

系统中存在数据库,为了方便对数据库中的表格进行读取,可以对每个表格创建相对应的类,类中的成员变量对应表格中相应的字段,成员方法可以字段进行读取和写入。这种封装了数据表所有字段的类我们称为实体类,这些实体类可以通过一些方法进行数据的操作,而包含这些方法的类,我们成为数据库操作类。本系统分为 9 个实体类,分别是 Back 类、BookInfo 类、BookType 类、Borrow 类、Operater 类、Order 类、OrderAndBookInfo 类、Reader 类、User 类。

1. Back 类

Back 类对表 tb_bookInfo 和 表 tb_reader 进行操作,关键代码如下:

```java
public class Back {
    private String bookISBN;
    private String bookname;
    private String operatorId;
    private String borrowDate;
    private String backDate;
```

```java
    private String readerName;
    private String readerISBN;
    private int typeId;
    private int id;
    public int getId() {
        return id;
    }
    public void setId(int id) {
        this.id = id;
    }
    public int getTypeId() {
        return typeId;
    }
    public void setTypeId(int typeId) {
        this.typeId = typeId;
    }
    public String getBackDate() {
        return backDate;
    }
    public void setBackDate(String backDate) {
        this.backDate = backDate;
    }
    public String getBookISBN() {
        return bookISBN;
    }
    public void setBookISBN(String bookISBN) {
        this.bookISBN = bookISBN;
    }
    public String getBookname() {
        return bookname;
    }
    public void setBookname(String bookname) {
        this.bookname = bookname;
    }
    public String getBorrowDate() {
        return borrowDate;
    }
    public void setBorrowDate(String borrowDate) {
        this.borrowDate = borrowDate;
```

```java
    }
    public String getOperatorId() {
        return operatorId;
    }
    public void setOperatorId(String operatorId) {
        this.operatorId = operatorId;
    }
    public String getReaderISBN() {
        return readerISBN;
    }
    public void setReaderISBN(String readerISBN) {
        this.readerISBN = readerISBN;
    }
    public String getReaderName() {
        return readerName;
    }
    public void setReaderName(String readerName) {
        this.readerName = readerName;
    }
}
```

2. BookInfo 类

BookInfo 类对表 tb_bookInfo 进行操作,关键代码如下:

```java
public class BookInfo {
    private String ISBN;
    private String typeid;
    private String writer;
    private String translator;
    private String publisher;
    private Date date;
    private Double price;
    private String bookname;
    public String getBookname() {
        return bookname;
    }
    public void setBookname(String bookname) {
        this.bookname = bookname;
    }
    public Date getDate() {
        return date;
```

```java
    }
    public void setDate(Date date) {
        this.date = date;
    }
    public String getISBN() {
        return ISBN;
    }
    public void setISBN(String isbn) {
        ISBN = isbn;
    }
    public Double getPrice() {
        return price;
    }
    public void setPrice(Double price) {
        this.price = price;
    }
    public String getPublisher() {
        return publisher;
    }
    public void setPublisher(String publisher) {
        this.publisher = publisher;
    }
    public String getTranslator() {
        return translator;
    }
    public void setTranslator(String translator) {
        this.translator = translator;
    }
    public String getTypeid() {
        return typeid;
    }
    public void setTypeid(String typeid) {
        this.typeid = typeid;
    }
    public String getWriter() {
        return writer;
    }
    public void setWriter(String writer) {
        this.writer = writer;
```

        }
}

3. BookType 类

BookType 类对表 tb_bookType 进行操作,关键代码如下:

```java
public class BookType {
    private String id;
    private String typeName;
    private String days;
    private String fk;
    public String getFk() {
        return fk;
    }
    public void setFk(String fk) {
        this.fk = fk;
    }
    public String getDays() {
        return days;
    }
    public void setDays(String days) {
        this.days = days;
    }
    public String getId() {
        return id;
    }
    public void setId(String id) {
        this.id = id;
    }
    public String getTypeName() {
        return typeName;
    }
    public void setTypeName(String typeName) {
        this.typeName = typeName;
    }
}
```

4. Borrow 类

Borrow 类对表 tb_borrow 进行操作,关键代码如下:

```java
public class Borrow {
    private int id;
    private String bookISBN;
```

```java
private String readerISBN;
private String num;
private String borrowDate;
private String backDate;
private String bookName;
public String getBookName() {
    return bookName;
}
public void setBookName(String bookName) {
    this.bookName = bookName;
}
public String getBackDate() {
    return backDate;
}
public void setBackDate(String backDate) {
    this.backDate = backDate;
}
public String getBookISBN() {
    return bookISBN;
}
public void setBookISBN(String bookISBN) {
    this.bookISBN = bookISBN;
}
public String getBorrowDate() {
    return borrowDate;
}
public void setBorrowDate(String borrowDate) {
    this.borrowDate = borrowDate;
}
public String getNum() {
    return num;
}
public void setNum(String num) {
    this.num = num;
}
public String getReaderISBN() {
    return readerISBN;
}
public void setReaderISBN(String readerISBN) {
```

```java
        this.readerISBN = readerISBN;
    }
    public int getId() {
        return id;
    }
    public void setId(int id) {
        this.id = id;
    }
}
```

5. Operater 类

Operater 类对表 tb_operator 进行操作,关键代码如下:

```java
public class Operater {
    private String id;
    private String name;
    private String grade;
    private String password;
    public String getGrade() {
        return grade;
    }
    public void setGrade(String grade) {
        this.grade = grade;
    }
    public String getId() {
        return id;
    }
    public void setId(String id) {
        this.id = id;
    }
    public String getName() {
        return name;
    }
    public void setName(String name) {
        this.name = name;
    }
    public String getPassword() {
        return password;
    }
    public void setPassword(String password) {
        this.password = password;
```

        }
}

6. Order 类

Order 类对表 tb_order 进行操作，关键代码如下：

```java
public class Order {
    private String ISBN;
    private Date date;
    private String number;
    private String operator;
    private String checkAndAccept;
    private String zk;
    public String getCheckAndAccept() {
        return checkAndAccept;
    }
    public void setCheckAndAccept(String checkAndAccept) {
        this.checkAndAccept = checkAndAccept;
    }
    public Date getDate() {
        return date;
    }
    public void setDate(Date date) {
        this.date = date;
    }
    public String getISBN() {
        return ISBN;
    }
    public void setISBN(String isbn) {
        ISBN = isbn;
    }
    public String getNumber() {
        return number;
    }
    public void setNumber(String number) {
        this.number = number;
    }
    public String getOperator() {
        return operator;
    }
    public void setOperator(String operator) {
```

```java
        this.operator = operator;
    }
    public String getZk() {
        return zk;
    }
    public void setZk(String zk) {
        this.zk = zk;
    }
}
```

由于篇幅有限，我们只列举了6个实体类，其他三个实体类构造方式相近。

### 9.5.4 数据库操作类的实现

纯Java驱动是有JDBC驱动直接访问数据库的，由于完全用Java语言编写，所以具有跨平台的功能，并且效率高。本书使用的Mircosoft SQL Server 2000数据库SP4版，先要下载驱动程序msbase.jar、mssqlserver.jar和msutil.jar三个jar包，然后把jar包加载到eclipse开发环境中。如果使用的是Mircosoft SQL Server 2005数据库，要下载驱动程序sqljdbc.jar包。最后开发环境中还要加载hibernate3.zip包。

(1) 本系统的数据库操作类是Dao类，我们先要进行数据库的连接，要连接的数据库是db_library，SQL Server用户名是sa，密码是sa。关键代码如下：

```java
public class Dao {
    protected static String dbClassName = "com.microsoft.jdbc.sqlserver.SQLServerDriver";
    protected static String dbUrl = "jdbc:sqlserver://localhost:1433;integratedSecurity=true;dataBaseName=db_library";
    protected static String dbUser = "sa";
    protected static String dbPwd = "sa";
    protected static String second = null;
    private static Connection conn = null;
    private Dao() {
        try {
            if (conn == null) {
                Class.forName(dbClassName).newInstance();
                conn = DriverManager.getConnection(dbUrl, dbUser, dbPwd);
            }
            else
                return;
        } catch (Exception ee) {
            ee.printStackTrace();
```

    }
}

(2) 建立连接后,就可以通过 SQL 语句对数据库进行操作了。

```java
private static ResultSet executeQuery(String sql) {
    try {
        if(conn = = null)
            new Dao();
        return conn.createStatement(ResultSet.TYPE_SCROLL_SENSITIVE,
        ResultSet.CONCUR_UPDATABLE).executeQuery(sql);
    } catch (SQLException e) {
        e.printStackTrace();
        return null;
    } finally {
    }
}
private static int executeUpdate(String sql) {
    try {
        if(conn = = null)
            new Dao();
        return conn.createStatement().executeUpdate(sql);
    } catch (SQLException e) {
        System.out.println(e.getMessage());
        return -1;
    } finally {
    }
}
```

(3) 访问完数据库,应该关闭数据库,释放资源。

```java
public static void close() {
    try {
        conn.close();
    } catch (SQLException e) {
        e.printStackTrace();
    }finally{
        conn = null;
    }
}
```

(4) 由于篇幅有限,我们以对读者信息表执行的相关操作为例,来说明数据库操作类的使用。关键代码如下:

//添加读者信息

```java
public static int InsertReader (String name, String sex, String age, String identityCard, Date date, String maxNum, String tel, Double keepMoney, String zj, String zy, Date bztime, String ISBN){
    int i = 0;
    try{
        String sql = "insert into tb_reader(name,sex,age,identityCard,date,maxNum,tel,keepMoney,zj,zy,bztime,ISBN) values('" + name + "','" + sex + "','" + age + "','" + identityCard + "','" + date + "','" + maxNum + "','" + tel + "'," + keepMoney + ",'" + zj + "','" + zy + "','" + bztime + "','" + ISBN + "')";
        System.out.println(sql);
        i = Dao.executeUpdate(sql);
    }catch(Exception e){
        e.printStackTrace();
    }
    Dao.close();
    return i;
}

//查询读者信息
public static List selectReader() {
    List list = new ArrayList();
    String sql = "select * from tb_reader";
    ResultSet rs = Dao.executeQuery(sql);
    try {
        while (rs.next()) {
            Reader reader = new Reader();
            reader.setName(rs.getString("name"));
            reader.setSex(rs.getString("sex"));
            reader.setAge(rs.getString("age"));
            reader.setIdentityCard(rs.getString("identityCard"));
            reader.setDate(rs.getDate("date"));
            reader.setMaxNum(rs.getString("maxNum"));
            reader.setTel(rs.getString("tel"));
            reader.setKeepMoney(rs.getDouble("keepMoney"));
            reader.setZj(rs.getInt("zj"));
            reader.setZy(rs.getString("zy"));
            reader.setISBN(rs.getString("ISBN"));
            reader.setBztime(rs.getDate("bztime"));
            list.add(reader);
```

```java
            }
        } catch (Exception e) {
            e.printStackTrace();
        }
        Dao.close();
        return list;
    }

    public static List selectReader(String readerISBN) {
        List list = new ArrayList();
        String sql = "select *  from tb_reader where ISBN = '" + readerISBN + "'";
        ResultSet rs = Dao.executeQuery(sql);
        try {
            while (rs.next()) {
                Reader reader = new Reader();
                reader.setName(rs.getString("name"));
                reader.setSex(rs.getString("sex"));
                reader.setAge(rs.getString("age"));
                reader.setIdentityCard(rs.getString("identityCard"));
                reader.setDate(rs.getDate("date"));
                reader.setMaxNum(rs.getString("maxNum"));
                reader.setTel(rs.getString("tel"));
                reader.setKeepMoney(rs.getDouble("keepMoney"));
                reader.setZj(rs.getInt("zj"));
                reader.setZy(rs.getString("zy"));
                reader.setISBN(rs.getString("ISBN"));
                reader.setBztime(rs.getDate("bztime"));
                list.add(reader);
            }
        } catch (Exception e) {
            e.printStackTrace();
        }
        Dao.close();
        return list;
    }
    //修改读者信息
    public static int UpdateReader(String id, String name, String sex, String age, String identityCard, Date date, String maxNum, String tel, Double keepMoney, String zj, String zy, Date bztime, String ISBN){
        int i = 0;
```

```
    try{
        String sql = "update tb_reader set name = '" + name + "',sex = '" + sex + "',age = '" +
        age + "',identityCard = '" + identityCard + "',date = '" + date + "',maxNum = '" +
        maxNum + "',tel = '" + tel + "',keepMoney = " + keepMoney + ",zj = '" + zj + "',zy
        = '" + zy + "',bztime = '" + bztime + "'where ISBN = '" + ISBN + "'";
        i = Dao.executeUpdate(sql);
    }catch(Exception e){
        e.printStackTrace();
    }
    Dao.close();
    return i;
}
//删除读者信息
public static int DelReader(String ISBN){
    int i = 0;
    try{
        String sql = "delete from tb_reader where ISBN = '" + ISBN + "'";
        i = Dao.executeUpdate(sql);
    }catch(Exception e){
        e.printStackTrace();
    }
    Dao.close();
    return i;
}
```

### 9.5.5 读者信息管理模块的实现

读者信息管理模块分为读者信息添加模块和读者信息修改与删除模块。

#### 9.5.5.1 读者信息添加模块

**关键步骤与代码**

1. 设计读者信息添加模块窗口

(1) 定义一个 ReaderAddIFrame 类,继承 JInternalFrame 框架类。

(2) 窗口中主要包含的组件:5 个 JTextField 控件,2 个 JComboBox 控件,1 个 JFormattedTextField 控件,2 个 JButton 控件。

(3) 窗口的布局方式:采用边框布局方式,网格布局方式。

设计窗口如图 9.3 所示。

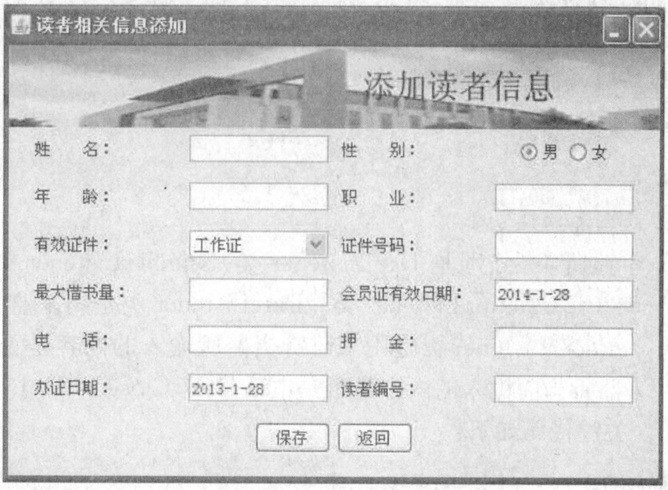

实验 9 综合项目 2——图书馆信息管理系统

图 9.3 添加读者信息

在构造方法中添加如下关键代码：

```java
public ReaderAddIFrame() {
    super();
    setTitle("读者相关信息添加");
    setIconifiable(true);
    setClosable(true);
    setBounds(100, 100, 500, 350);
    final JLabel logoLabel = new JLabel();
    ImageIcon readerAddIcon = CreatecdIcon.add("readerAdd.jpg");
    logoLabel.setIcon(readerAddIcon);
    logoLabel.setOpaque(true);
    logoLabel.setBackground(Color.CYAN);
    logoLabel.setPreferredSize(new Dimension(400, 60));
    getContentPane().add(logoLabel, BorderLayout.NORTH);
    final JPanel panel = new JPanel();
    panel.setLayout(new FlowLayout());
    getContentPane().add(panel);
    final JPanel panel_1 = new JPanel();
    final GridLayout gridLayout = new GridLayout(0, 4);
    gridLayout.setVgap(15);
    gridLayout.setHgap(10);
    panel_1.setLayout(gridLayout);
    panel_1.setPreferredSize(new Dimension(450, 200));
    panel.add(panel_1);
    final JLabel label_2 = new JLabel();
    label_2.setText("姓    名:");
```

```
        panel_1.add(label_2);
        readername = new JTextField();
        readername.setDocument(new MyDocument(10));
        panel_1.add(readername);
}
```

2. 实现读者信息添加模块功能

类中定义了 4 个监听类，分别是 DateListener 类、NumberListener 类、TelListener 类、ButtonAddListener 类、CloseActionListener 类。DateListener 类是用来监听当前所输入的内容是否是时间格式，NumberListener 类是用来监听当前所输入的内容是否是数字或退格键，TelListener 类是用来监听当前输入的内容是否是电话号码，CloseActionListener 类是用来监听窗口是否被关闭。关键代码如下：

```
class DateListener extends KeyAdapter {
    public void keyTyped(KeyEvent e) {
        if(bztime.getText().isEmpty()){
            JOptionPane.showMessageDialog(null, "时间格式请使用\"2007 - 05 - 10\"格式");
        }
    }
}
class NumberListener extends KeyAdapter {
    public void keyTyped(KeyEvent e) {
        String numStr = "0123456789" + (char)8;
        if(numStr.indexOf(e.getKeyChar())<0){
            e.consume();
        }
    }
}
class ButtonAddListener implements ActionListener {
    private final JRadioButton button1;
    ButtonAddListener(JRadioButton button1) {
        this.button1 = button1;
    }
    public void actionPerformed(final ActionEvent e) {
        if(readername.getText().length() = = 0){
            JOptionPane.showMessageDialog(null, "读者姓名文本框不可为空");
            return;
        }
        ......
        String sex = "1";
```

```
                if(!button1.isSelected()){
                    sex = "2";}
                String zj = String.valueOf(comboBox.getSelectedIndex());
                System.out.println(comboBox.getSelectedIndex());
                i = Dao.InsertReader(readername.getText().trim(), sex.trim(),
                 age.getText().trim(), zjnumber.getText().trim(), Date.valueOf(date.
                 getText().trim()), maxnumber.getText().trim(),tel.getText().trim(),
                Double.valueOf(keepmoney.getText().trim()),zj,zy.getText().trim(),Date.
                valueOf(bztime.getText().trim()),ISBN.getText().trim());
                System.out.println(i);
                if(i = = 1){
                    JOptionPane.showMessageDialog(null, "添加成功!");
                    doDefaultCloseAction();
                }
            }
        }
    }
    class TelListener extends KeyAdapter {
        public void keyTyped(KeyEvent e) {
            String numStr = "0123456789 - " + (char)8;
            if(numStr.indexOf(e.getKeyChar())<0){
                e.consume();
            }
        }
    }
// 添加关闭按钮的事件监听器
    class CloseActionListener implements ActionListener {
            public void actionPerformed(final ActionEvent e) {
                doDefaultCloseAction();
            }
        }
    }
}
```

#### 9.5.5.2 读者信息修改与删除模块

**关键步骤与代码**

1. 设计读者信息修改与删除模块窗口

(1) 定义一个 ReaderModiAndDelIFrame 类,继承 JInternalFrame 框架类。

(2) 窗口中主要包含的组件:10 个 JTextField 控件,1 个 JTable 控件,1 个 JComboBox 控件,1 个 ButtonGroup 控件,2 个 JRadioButton 控件。

(3) 窗口的布局方式:采用边框布局方式,流式布局方式。

设计窗口如图 9.4 所示。

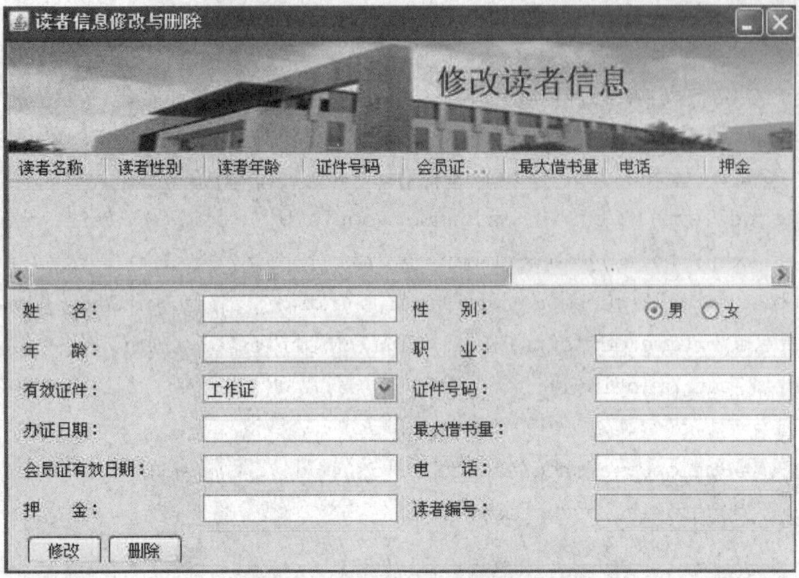

图9.4 读者信息修改与删除

通过循环读取读者信息，保存在二维数组 results 中。表格显示在一维数组 columnNames 中存储的列名。在类中添加如下关键代码：

```java
public class ReaderModiAndDelIFrame extends JInternalFrame {
    private JTextField keepmoney;
    private ButtonGroup buttonGroup = new ButtonGroup();
    private JTable table;
    private JTextField ISBN;
    private JTextField zy;
    private JTextField tel;
    private JTextField date;
    private JTextField maxnumber;
    private JTextField bztime;
    private JTextField zjnumber;
    private JComboBox comboBox;
    private JTextField age;
    private JTextField readername;
    private JRadioButton JRadioButton1;
    private JRadioButton JRadioButton2;
    private String[] columnNames = { "读者名称", "读者性别", "读者年龄", "证件号码", "会员证有效日期","最大借书量", "电话","押金","证件","职业","读者编号","读者办证时间" };
    private String[] array = new String[]{"身份证","军人证","学生证","工作证"};
    String id;
```

```java
//读取读者信息
private Object[][] getFileStates(List list){
    Object[][]results = new Object[list.size()][columnNames.length];
    for(int i = 0;i<list.size();i++){
        Reader reader = (Reader)list.get(i);
        results[i][0] = reader.getName();
        String sex;
        if(reader.getSex().equals("1")){
            sex = "男";
        }
        else
            sex = "女";
        results[i][1] = sex;
        results[i][2] = reader.getAge();
        results[i][3] = reader.getIdentityCard();
        results[i][4] = reader.getDate();
        results[i][5] = reader.getMaxNum();
        results[i][6] = reader.getTel();
        results[i][7] = reader.getKeepMoney();
        results[i][8] = array[reader.getZj()];
        results[i][9] = reader.getZy();
        results[i][10] = reader.getISBN();
        results[i][11] = reader.getBztime();
    }
    return results;
}
//构造方法初始化界面
public ReaderModiAndDelIFrame() {
    super();
    setIconifiable(true);
    setClosable(true);
    setTitle("读者信息修改与删除");
    setBounds(100, 100, 600, 420);
    final JPanel panel = new JPanel();
    panel.setLayout(new BorderLayout());
    panel.setPreferredSize(new Dimension(400, 80));
    getContentPane().add(panel, BorderLayout.NORTH);
    final JLabel logoLabel = new JLabel();
    ImageIcon readerModiAndDelIcon = CreatecdIcon.add("readerModiAndDel.jpg");
```

```java
logoLabel.setIcon(readerModiAndDelIcon);
logoLabel.setBackground(Color.CYAN);
logoLabel.setOpaque(true);
logoLabel.setPreferredSize(new Dimension(400, 80));
panel.add(logoLabel);
logoLabel.setText("读者信息修改 logo(400 * 80)");
final JPanel panel_1 = new JPanel();
panel_1.setLayout(new BorderLayout());
getContentPane().add(panel_1);
final JScrollPane scrollPane = new JScrollPane();
scrollPane.setPreferredSize(new Dimension(0, 100));
panel_1.add(scrollPane, BorderLayout.NORTH);
final DefaultTableModel model = new DefaultTableModel();
Object[][] results = getFileStates(Dao.selectReader());
model.setDataVector(results, columnNames);
table = new JTable();
table.setModel(model);
scrollPane.setViewportView(table);
table.setAutoResizeMode(JTable.AUTO_RESIZE_OFF);
table.addMouseListener(new TableListener());
final JPanel panel_2 = new JPanel();
final GridLayout gridLayout = new GridLayout(0, 4);
gridLayout.setVgap(9);
panel_2.setLayout(gridLayout);
panel_2.setPreferredSize(new Dimension(0, 200));
panel_1.add(panel_2, BorderLayout.SOUTH);
final JLabel label_1 = new JLabel();
label_1.setText(" 姓    名:");
panel_2.add(label_1);
readername = new JTextField();
readername.setDocument(new MyDocument(10));
panel_2.add(readername);
    ......
final JPanel panel_4 = new JPanel();
panel_4.setMaximumSize(new Dimension(0, 0));
final FlowLayout flowLayout = new FlowLayout();
flowLayout.setVgap(0);
flowLayout.setHgap(4);
panel_4.setLayout(flowLayout);
```

```
    panel_2.add(panel_4);
    final JButton button = new JButton();
    button.setHorizontalTextPosition(SwingConstants.CENTER);
    panel_4.add(button);
    button.setText("修改");
    button.addActionListener(new ModiButtonListener(model));
    final JButton buttonDel = new JButton();
    panel_4.add(buttonDel);
    buttonDel.setText("删除");
    buttonDel.addActionListener(new DelButtonListener(model));
    setVisible(true);
}
```

2. 实现读者信息修改与删除模块功能

类中定义了 6 个监听类，分别是 TableListener 类、NumberListener 类、TelListener 类、DelButtonListener 类、ModiButtonListener 类、KeepmoneyListener 类。TableListener 类是用来监听鼠标单击事件，NumberListener 类是用来监听当前所输入的内容是否是数字或退格键，TelListener 类是用来监听当前输入的内容是否是电话号码，DelButtonListener 类是用来监听删除按钮发生的事件。ModiButtonListener 类是用来监听修改按钮发生的事。KeepmoneyListener 类是用来监听当前所输入的内容是否是有数字和退格键。关键代码如下：

```
class TableListener extends MouseAdapter {
    public void mouseClicked(final MouseEvent e) {
        int selRow = table.getSelectedRow();
        readername.setText(table.getValueAt(selRow, 0).toString().trim());
        if(table.getValueAt(selRow, 1).toString().trim().equals("男"))
            JRadioButton1.setSelected(true);
        else
            JRadioButton2.setSelected(true);
        age.setText(table.getValueAt(selRow, 2).toString().trim());
        zjnumber.setText(table.getValueAt(selRow, 3).toString().trim());
        date.setText(table.getValueAt(selRow, 4).toString().trim());
        maxnumber.setText(table.getValueAt(selRow, 5).toString().trim());
        tel.setText(table.getValueAt(selRow, 6).toString().trim());
        keepmoney.setText(table.getValueAt(selRow, 7).toString().trim());
        comboBox.setSelectedItem(table.getValueAt(selRow, 8).toString().trim());
        zy.setText(table.getValueAt(selRow, 9).toString().trim());
        ISBN.setText(table.getValueAt(selRow, 10).toString().trim());
        bztime.setText(table.getValueAt(selRow, 11).toString().trim());
    }
```

```java
    }
    final class NumberListener extends KeyAdapter {
        public void keyTyped(KeyEvent e) {
            String numStr = "0123456789" + (char)8;
            if(numStr.indexOf(e.getKeyChar())<0){
                e.consume();
            }
        }
    }
    private final class DelButtonListener implements ActionListener {
        private final DefaultTableModel model;
        private DelButtonListener(DefaultTableModel model) {
            this.model = model;
        }
        public void actionPerformed(final ActionEvent e) {
            int i = Dao.DelReader(ISBN.getText().trim());
            if(i = = 1){
                JOptionPane.showMessageDialog(null,"删除成功");
                Object[][] results = getFileStates(Dao.selectReader());
                model.setDataVector(results,columnNames);
                table.setModel(model);
            }
        }
    }
    class ModiButtonListener implements ActionListener {
        private final DefaultTableModel model;
        ModiButtonListener(DefaultTableModel model) {
            this.model = model;
        }
        public void actionPerformed(final ActionEvent e) {
            if(readername.getText().length() = = 0){
                JOptionPane.showMessageDialog(null,"读者姓名文本框不可为空");
                return;
            }
            ……
            String sex = "1";
            if(!JRadioButton1.isSelected()){
                sex = "2";}
            String zj = String.valueOf(comboBox.getSelectedIndex());
```

```java
                System.out.println(comboBox.getSelectedIndex());
                int i = Dao.UpdateReader(id, readername.getText().trim(), sex, age.
                getText().trim(), zjnumber.getText().trim(), Date.valueOf(date.
                getText().trim()), maxnumber.getText().trim(), tel.getText().trim(),
                Double.valueOf(keepmoney.getText().trim()), zj, zy.getText().trim(),
                Date.valueOf(bztime.getText().trim()), ISBN.getText().trim());
                System.out.println(i);
                if(i==1){
                    JOptionPane.showMessageDialog(null,"修改成功");
                    Object[][] results = getFileStates(Dao.selectReader());
                    model.setDataVector(results,columnNames);
                    table.setModel(model);
                }
            }
        }
        class TelListener extends KeyAdapter {
            public void keyTyped(KeyEvent e) {
                String numStr = "0123456789-" + (char)8;
                if(numStr.indexOf(e.getKeyChar())<0){
                    e.consume();
                }
            }
        }
        class KeepmoneyListener extends KeyAdapter {
            public void keyTyped(KeyEvent e) {
                String numStr = "0123456789" + (char)8; //只允许输入数字与退格键
                if(numStr.indexOf(e.getKeyChar())<0){
                    e.consume();
                }
                if(keepmoney.getText().length()>2||keepmoney.getText().length()<0){
                    e.consume();
                }
            }
        }
}
```

### 9.5.6 图书类别管理模块的实现

图书类别信息管理模块分为图书类别添加模块和图书类别修改模块。

### 9.5.6.1 图书类别添加模块

**关键步骤与代码**

**设计图书类别添加模块窗口**

(1) 定义一个 BookTypeAddIFrame 类,继承 JInternalFrame 框架类。

(2) 窗口中主要包含的组件:2 个 JTextField 控件,1 个 JFormattedTextField 控件,2 个 JButton 控件。

(3) 窗口的布局方式:采用边框布局方式。

设计窗口如图 9.5 所示。

图 9.5 图书类别添加

因篇幅有限,代码不再赘述。

### 9.5.6.2 图书类别修改与删除模块

**关键步骤与代码**

1. 设计图书类别修改与删除模块窗口

(1) 定义一个 BookTypeModiAndDelIFrame 类,继承 JInternalFrame 框架类。

(2) 窗口中主要包含的组件:3 个 JTextField 控件,1 个 JComboBox 控件,1 个 JTable 控件,2 个 JButton 控件。

(3) 窗口的布局方式:采用边框布局方式。

设计窗口如图 9.6 所示。

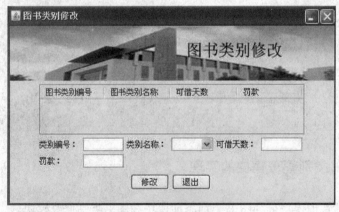

图 9.6 图书类别修改

2. 实现图书类别修改与删除模块功能

类中定义了3个监听类,分别是 TableListener 类、ButtonAddListener 类、CloseActionListener 类。TableListener 类是用来监听鼠标单击表格事件,ButtonAddListener 类是用来监听修改按钮的发生事件,CloseActionListener 类是用来监听退出按钮的发生事件。关键代码如下:

// 添加表格的事件监听器

```java
class TableListener extends MouseAdapter {
    public void mouseClicked(final MouseEvent e) {
        int selRow = table.getSelectedRow();
        BookTypeId.setText(table.getValueAt(selRow, 0).toString().trim());
        bookTypeModel.setSelectedItem(table.getValueAt(selRow, 1).toString().trim());
        days.setText(table.getValueAt(selRow, 2).toString().trim());
        fk.setText(table.getValueAt(selRow, 3).toString().trim());

    }
}
```

// 添加修改按钮的事件监听器

```java
class ButtonAddListener implements ActionListener{
    public void actionPerformed(ActionEvent e){
        Object selectedItem = bookTypeModel.getSelectedItem();
        int i = Dao.UpdatebookType(BookTypeId.getText().trim(),selectedItem.toString(),
        days.getText().trim(),fk.getText().trim());
        System.out.println(i);
        if(i==1){
            JOptionPane.showMessageDialog(null,"修改成功");
            Object[][] results = getFileStates(Dao.selectBookCategory());
            model.setDataVector(results,columnNames);
            table.setModel(model);
        }
    }
}
```

// 添加退出按钮的事件监听器

```java
class CloseActionListener implements ActionListener {
    public void actionPerformed(final ActionEvent e) {
        doDefaultCloseAction();
    }
}
```

## 9.5.7 图书信息管理模块的实现

图书信息管理模块分为图书信息添加模块和图书信息修改模块。

### 9.5.7.1 图书信息添加模块

**关键步骤与代码**

1. 设计图书信息添加模块窗口

(1) 定义一个 BookAddIFrame 类，继承 JInternalFrame 框架类。

(2) 窗口中主要包含的组件:5 个 JTextField 控件,1 个 JFormattedTextField 控件,2 个 JComboBox 控件,2 个 JButton 控件。

(3) 窗口的布局方式:采用边框布局方式、网格布局方式。

设计窗口如图 9.7 所示。

图 9.7 图书信息添加

2. 实现图书信息添加模块功能

类中定义了 5 个监听类,分别是 ISBNFocusListener 类、CloseActionListener 类、addBookActionListener 类、NumberListener 类。ISBNFocusListener 类是用来监听用户正在输入的图书编号是否存在于 bookinfo 表中,如果存在,提示警告信息,以此来保证图书编号的唯一。addBookActionListener 类是用来监听添加按钮的发生事件。CloseActionListener 类是用来监听关闭按钮的发生事件。NumberListener 类是用来监听用户正在输入的是否是数字。关键代码如下:

//添加查询图书编号的事件监听器

```
class ISBNFocusListener extends FocusAdapter {
    public void focusLost(FocusEvent e){
        if(!Dao.selectBookInfo(ISBN.getText().trim()).isEmpty()){
            JOptionPane.showMessageDialog(null, "添加书号重复!");
            return;
        }
    }
}
```

// 添加关闭按钮的事件监听器

```java
class CloseActionListener implements ActionListener {
    public void actionPerformed(final ActionEvent e) {
        doDefaultCloseAction();
    }
}
    // 添加按钮的单击事件监听器
class addBookActionListener implements ActionListener {
    public void actionPerformed(final ActionEvent e) {
            if(ISBN.getText().length() == 0){
            JOptionPane.showMessageDialog(null,"书号文本框不可以为空");
            return;
        }
            if(ISBN.getText().length() != 13){
            JOptionPane.showMessageDialog(null,"书号文本框输入位数为 13 位");
            return;
        }
            if(bookName.getText().length() == 0){
            JOptionPane.showMessageDialog(null,"图书名称文本框不可以为空");
            return;
        }
            if(writer.getText().length() == 0){
            JOptionPane.showMessageDialog(null,"作者文本框不可以为空");
            return;
        }
            if(pubDate.getText().length() == 0){
            JOptionPane.showMessageDialog(null,"出版日期文本框不可以为空");
            return;
        }
            if(price.getText().length() == 0){
            JOptionPane.showMessageDialog(null,"单价文本框不可以为空");
            return;
        }
        String ISBNs = ISBN.getText().trim();
        Object selectedItem = bookType.getSelectedItem();
        if (selectedItem == null)
            return;
        Item item = (Item) selectedItem;
        String bookTypes = item.getId();
        String translators = translator.getText().trim();
        String bookNames = bookName.getText().trim();
```

```
            String writers = writer.getText().trim();
        String publishers = (String)publisher.getSelectedItem();
        String pubDates = pubDate.getText().trim();
        String prices = price.getText().trim();
            int i = Dao.Insertbook(ISBNs, bookTypes, bookNames, writers, translators,
            publishers,java.sql.Date.valueOf(pubDates),Double.parseDouble(prices));
            if(i= =1){
                JOptionPane.showMessageDialog(null,"添加成功");
                doDefaultCloseAction();
            }
        }
    }
}
class NumberListener extends KeyAdapter {
    public void keyTyped(KeyEvent e) {
        String numStr = "0123456789." + (char)8;
        if(numStr.indexOf(e.getKeyChar())<0){
            e.consume();
        }
    }
}
```

#### 9.5.7.2 图书信息修改模块

**关键步骤与代码**

1. 设计图书信息修改模块窗口

(1) 定义一个 BookModiAndDelIFrame 类,继承 JInternalFrame 框架类。

(2) 窗口中主要包含的组件:5 个 JTextField 控件,2 个 JFormattedTextField 控件,1 个 JComboBox 控件,1 个 JTable 控件。

(3) 窗口的布局方式:采用边框布局方式、流式布局方式。

设计窗口如图 9.8 所示。

图 9.8 图书信息修改

2. 实现图书信息修改模块功能

类中定义了 3 个监听类,分别是 TableListener 类、addBookActionListener 类和 NumberListener 类。TableListener 类是用来监听鼠标单击表格事件。addBookActionListener 类是用来监听添加按钮的发生事件。NumberListener 类是用来监听用户正在输入的是否是数字。

### 9.5.8 新书管理模块的实现

新书管理模块分为新书订购模块和验证新书模块。

#### 9.5.8.1 新书管理模块

**关键步骤与代码**

1. 设计新书管理模块窗口

(1) 定义一个 newBookOrderIFrame 类,继承 JInternalFrame 框架类。

(2) 窗口中主要包含的组件:6 个 JTextField 控件,1 个 JFormattedTextField 控件,2 个 JComboBox 控件,1 个 ButtonGroup 控件,2 个 JRadioButton 控件。

(3) 窗口的布局方式:采用流式布局方式、网格布局方式。

设计窗口如图 9.9 所示。

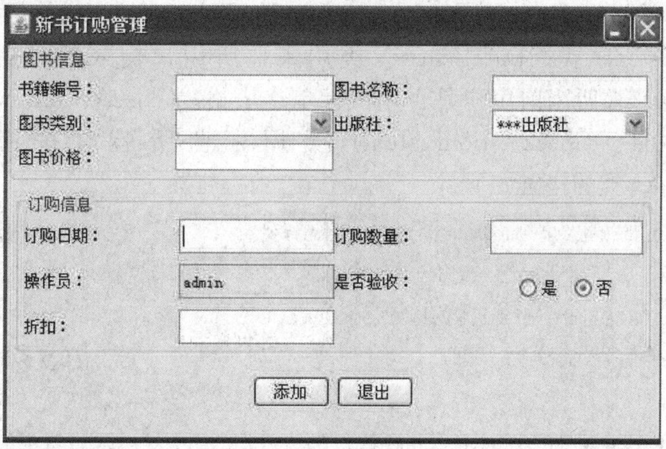

图 9.9 新书定购

2. 实现新书订购模块功能

类中定义了 6 个监听类,分别是 ButtonAddLisenter 类、DateListener 类、ISBNListener 类、NumberListener 类、ISBNListenerlostFocus 类、CloseActionListener 类。ButtonAddLisenter 类用来监听添加按钮的发生事件。DateListener 类用来监听当前用户输入的是否是合法的时间格式。ISBNListener 类是否使用回车键进行触发事件。NumberListener 类用来监听是否输入数字。

#### 9.5.8.2 验证新书模块

**关键步骤与代码**

1. 设计验证新书模块窗口

(1) 定义一个 newBookCheckIFrame 类,继承 JInternalFrame 框架类。

(2) 窗口中主要包含的组件:7 个 JTextField 控件,1 个 JFormattedTextField 控件,1 个 ButtonGroup 控件,2 个 JRadioButton 控件。

(3) 窗口的布局方式:采用网格布局方式。

设计窗口如图 9.10 所示。

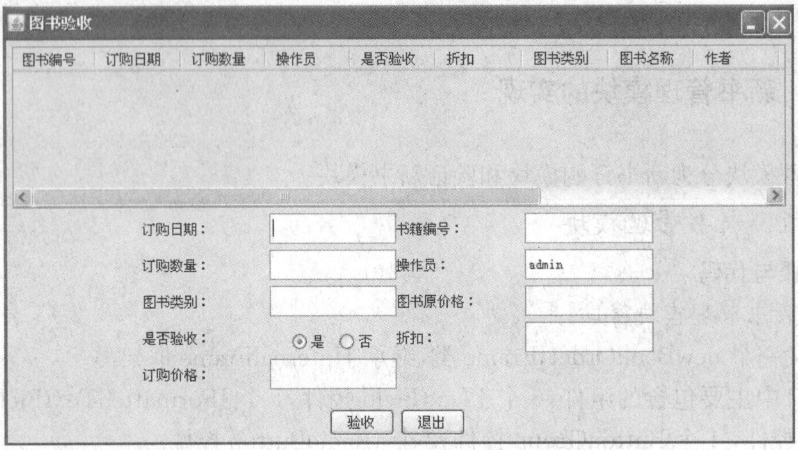

图 9.10 验证新书

2. 实现验证新书模块功能

类中定义了 4 个监听类,分别是 DateListener 类、TableListener 类、CloseActionListener 类、CheckActionListener 类。DateListener 类用来监听用户输入的日期格式是否合法。TableListener 类用来监听当前用户是否用鼠标单击表格。CloseActionListener 类监听退出按钮的发生事件。CheckActionListener 类用来监听验收按钮的发生事件。CheckActionListener 类的代码如下:

```java
class CheckActionListener implements ActionListener{
    private final DefaultTableModel model;
    CheckActionListener(DefaultTableModel model) {
        this.model = model;
    }
    public void actionPerformed(final ActionEvent e) {
        if(radioButton2.isSelected()){
            String ISBNs = ISBN.getText();
            int i = Dao.UpdateCheckBookOrder(ISBNs);
            if(i = = 1){
                JOptionPane.showMessageDialog(null, "验收成功!");
                Object[][] results = getFileStates(Dao.selectBookOrder());
                model.setDataVector(results,columnNames);
                table.setModel(model);
                radioButton1.setSelected(true);
            }
        }
```